An Invitation

to

Environmental Philosophy

An Invitation
to
Environmental
Philosophy

Editor

Anthony Weston

Contributors

David Abram

Jim Cheney

Val Plumwood

Holmes Rolston, III

Anthony Weston

New York   Oxford

Oxford University Press

1999

Oxford University Press

Oxford   New York
Athens   Auckland   Bangkok   Bogotá   Buenos Aires   Calcutta
Cape Town   Chennai   Dar es Salaam   Delhi   Florence   Hong Kong   Istanbul
Karachi   Kuala Lumpur   Madrid   Melbourne   Mexico City   Mumbai
Nairobi   Paris   São Paulo   Singapore   Taipei   Tokyo   Toronto   Warsaw

and associated companies in
Berlin   Ibadan

Published by Oxford University Press, Inc.
198 Madison Avenue, New York, New York 10016

Oxford is a registered trademark of Oxford University Press

Library of Congress Cataloging-in-Publication Data

An invitation to environmental philosophy / edited by Anthony Weston ;
    contributors: David Abram . . . [et al.].
        p.   cm.
      ISBN 0-19-512203-8 (cloth : alk. paper). — ISBN 0-19-512204-6
    (pbk. : alk. paper)
      1. Environmental ethics.   I. Weston, Anthony, 1954–   .
    GE42.I59   1998   97–48817
    179'.1—dc21   CIP

GE 42 I59 1999
OCLC: 38090910
An invitation to
environmental philosophy

9 8 7 6 5 4 3 2 1

Printed in the United States of America
on acid-free paper

—•—

Contents

Introduction   1

The environmental crisis is a crisis of the senses, of imagination, and of our tools for thinking—our concepts and theories—themselves. Beyond recycling, beyond public policy, even beyond ecology, *philosophy* too can and must contribute to twenty-first–century environmentalism. In this book five well-known environmental philosophers offer five diverse but complementary "invitations" to philosophy in this new, vital, and profoundly suggestive key.

1. A More-Than-Human World, by David Abram   17

This book opens with a narrative of personal and sensory engagement with the "more-than-human" world (a term Abram coins here) in Indonesia and Nepal, ending with a return to Long Island, where the familiar dismissals of animal others and the world of the senses now appear strange and radically limited. How much more is possible than we usually assume!

2. Is It Too Late?, by Anthony Weston   43

How do things stand in our relation to nature—right now, as the Millennium turns? Disastrous, in some ways: the world is radically, heart-breakingly reduced from what it once was. On the other hand, to suppose that that there is nothing to be done sells the Earth short too. Who knows? Wild possibilities still abound, and right next to us too.

Can we value nature apart from ourselves—apart from humans? Certain philosophical arguments conclude that we cannot. In fact, however, we already do. A *human*-centered ethic is no more inevitable than a *self*-centered (or *male*-centered, or *European*-centered . . .) ethic, and recent struggles against these other "centrisms" can show us the way beyond human-centeredness too.

In an ever-widening series of circles the scope of ethics may expand beyond the human. We may begin to speak for other animals, rivers, mountains, even the whole Earth itself—even the *dirt.* But this conceptual terrain remains mostly a wild one. Rolston offers himself as a kind of wilderness guide to this new territory.

An environmental ethic is at the same time a way of understanding and a way of valuing—a practical track through the world—and therefore also a *story.* The nonhuman world is not merely the object of this knowing and this valuing, but takes *part* in it. Even the rocks, the sunrise, and the wildflowers invite what Indigenous peoples might call "sacramental practice."

The concluding section offers an extensive survey of environmental philosophy and many related fields, from eco-theology to permaculture, including suggestions for practical action, further reading, organizations to contact, and questions to keep on asking.

Acknowledgments

My thanks to Robert Miller of Oxford University Press, who first planted the seed that led to this volume, and supported it all the way to its full flowering. All of the contributors have been generous not only with their writing but with encouragement, suggestions, references, and advice. Other colleagues and friends were generous too, especially Tom Birch, Bob Jickling, Erin Malloy-Hanley, and Roger Gottlieb. Two students at Elon College, Sarah Nardotti and Jeff Horn, worked through initial drafts of the essays with me as an independent project in the Spring of 1997. I hope all will find their influence here, and their time well-spent.

Jan Davis produced the images that grace most of the essays' opening pages. Fran Cheney's sketch from the Waterville Prairie introduces Jim Cheney's essay, which ends back at that very prairie. And Linda Martindale's computer wizardry made long-distance electronic collaboration possible. My appreciation to all of you as well.

Remembering how deeply all beings intertwine, it is all life, finally, and this planet which sustains us all—the earth, the air, the fire, the water—to which we must give our largest thanks. May gratefulness ever grow on us.

Raleigh, North Carolina   A.W.
Spring Equinox, 1998

Contributors

David Abram is a magician and philosopher, and author of *The Spell of the Sensuous* (Pantheon, 1996), winner of the Lannan Prize for Best Non-Fiction of 1996.

Jim Cheney has published a series of influential articles in the journal *Environmental Ethics* and elsewhere. This essay was composed while at work with the Native Philosophy Project at Lakehead University in Thunder Bay, Ontario.

Val Plumwood is an Australian philosopher and bushwoman; author of *Feminism and the Mastery of Nature* (Routledge, 1993) and several forthcoming books, including *The Eye of the Crocodile*.

Holmes Rolston, III is Founding Associate Editor of the journal *Environmental Ethics* and author of many books in the field, including *Philosophy Gone Wild* (Prometheus, 1986) and *Conserving Natural Value* (Columbia University Press, 1994).

Anthony Weston teaches philosophy and global studies; is author of *Back to Earth: Tomorrow's Environmentalism* (Temple University Press, 1994) and other books; and is editor of this collection.

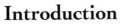

Introduction

I

We know that we face an environmental crisis. Pesticides in the groundwater, overfished oceans, ozone holes, soaring extinction rates: daily the news tells us more. The crisis is down the street too, as the last wooded spot on the block is bulldozed for a new shopping mall, the deer and woodcocks disappear, and you have to consult the newspaper to find out whether it is even safe to go outside and breathe.

If we ask *why* all of this is happening, we may get scientific answers. Certain pollutants are being added to the atmosphere (seven *billion* tons a year of carbon dioxide, for example). Fish populations need certain critical numbers to reproduce themselves, and so on. We may also get political answers. The privatization of property makes it hard to restrict ecologically disastrous uses of the land. Our media and politics are so money-driven that they tend to marginalize noncommercial interests, and so on again. There are historical answers, sociological answers (we learn to see ourselves as *distinct* from the world), religious and architectural answers as well.

But beneath these questions there is another kind of question too. We may also ask *why* again—*why* to these other answers in turn.

We see ourselves as distinct from the world, yes. When we speak of other animals we usually just say "animals," as if we are not animals too. But we can also ask why. *Why* is this way of talking, this way of thinking, so natural, even when we know perfectly well that we too are animals? Looked at philosophically, is it not even a little strange? (*More* than a little strange?) And what would an alternative look like? How can we now recognize and affirm our animality, rather than reject and deny it? How can we rejoin the community of life?

When we see ourselves as distinct from the world, we also open up the possibility of its exploitation and even destruction, even without a second thought. It doesn't seem to touch us. But again we find ourselves with deeper questions. Why, for example, do we believe that the ravaging of the land does not affect us? Considering that even an evening under the stars or a weekend camping can utterly reawaken our senses, what has the loss of a life *constantly* close to nature actually done to us? In our colleges and universities, students are supposed to come into maturity and claim the future in the absence of almost the whole of the biosphere: no plants, no wind (the windows don't even open, and in my school people don't even bother opening the *blinds*), no animals, not even any children. Is it any wonder that we feel alienated from nature, even betrayed by it?

These questions carry us deeper than a mere response to the current dangers—as vital as that response is too. An ancient Chinese saying holds that in every danger there is also opportunity. So what is the opportunity here? Beyond the crisis, beyond the fear that even our own lifeboat is springing leaks, what new opportunities does our growing environmental awareness open to us? The last twenty years have seen not only an awakening to the scope of the crisis, but also an explosion of interest in outdoor activities: backpacking and kayaking, hang-gliding and meditation. What does this new interest in nature mean? Has some deep sort of remembrance of nature been awakened here? Some sense of wild possibility? And if so, then the possibility—of *what*?

In short, we find ourselves with philosophical questions. Questions about the world of the senses and how it has changed. Questions about our feeling of—our insistence upon—human distinctness. Questions also about the imagination. Can we really imagine rejoining the rest of the world? What would it mean to visualize ourselves as one kind of being among a multitude of very different others on a co-inhabited Earth? How could we live out such a vision? Are there hints of such a vision already, right now, around us? Where?

II

It may be that our sense of danger overpowers any hope for opportunity down the road. Isn't it really too late? Aren't things already so bad that there is really nothing to be done, except maybe to enjoy what little is left before it too is gone? Certainly I find this attitude among my fellow outdoor enthusiasts, and it must be admitted that we are even accelerating the destruction ourselves—inadequately disposed human wastes now contaminate so much ground water in the wilderness, for instance, that water filters are necessary back-country equipment—though all of this still pales before the onslaught of bulldozers, chemicals, highways and the rest, also fueled in part by our own choices. That onslaught does not seem likely to stop. So can we do anything now to win more than a short reprieve before the inevitable ecological collapse?

Maybe not. Maybe the jig really is up. Apparently it will take millennia for the Earth to recover even from the damage we have caused already, if it can recover at all. Too many species are gone forever, or teeter on the brink of extinction. But there is another side to this story too. We are very far from understanding this Earth well enough to know what the real possibilities are. We don't even know how many species there are on this planet: maybe ten or a hundred or even a thousand times the number of species we know now, not that we truly "know" many of them either. Millions of unknown kinds of bugs remain, and even new species of large mammals keep turning up, one or two a year. A billion tons of that human-generated carbon dioxide disappears each year, we know not where, though this doesn't stop some people from predicting, or denying, global warming anyway. Science is uncovering an Earth far stranger than we thought—perhaps, in the end, stranger than we *can* think. So how can we think we know enough to declare the fate of the Earth sealed? Could it not be, instead, that the philosophical questions lead to a greater sense of complexity, caution—and surely hope too?

The environmental crisis is in part a crisis of concepts as well. Whenever the possibility of environmental ethics comes up, for example, someone always argues that truly valuing something beyond ourselves is not even possible. Mustn't we somehow always value in human terms (inescapably, they say) since after all we *are* humans, not elephants or rivers or mosquitoes? Is an environmental ethic—a nature-centered philosophy rather than a human-centered one—even conceivable?

Once again we must turn to philosophy, this time in a more analytic and conceptual key, for an answer. The alleged necessity of "human terms," human reference points, is much like the alleged necessity of the *self* as an ethical reference-point, in short like the alleged necessity of egoism ("isn't everything really self-interest?"). Only here the egoism in question is a kind of species-egoism. It is tempting to think that egoism in both senses is necessary, unavoidable. But in fact there is no such necessity. Afer all, we are all also, inescapably, *many* things: men or women, Americans or of other nationalities, as well as mammals, animals, life forms, earth beings. Could we not as well say that we must inevitably think as earth beings, since we *are* earth beings (too)? Or as animals? And in that case neither self-interest nor species-interest is the whole story for us. We are not so tightly tied to any particular "interest" as we might first have thought. A more intriguing question might be why we nonetheless think we are.

All of these philosophical questions are directly addressed by at least one of the essays in this book. We take them up and explore them and suggest some ways around and about. But we also need to insist that our aim in these essays is not to arrive at any kind of final *answer*. Instead, the book is an *invitation to the questions*. It is meant to open up possibilities. It uncovers the *depth* of the environmental crisis—understanding that crisis as a cultural and philosophical condition as well as a biological and social one—and in a way that both suggests directions for further thinking and also for practical action.

In fact we *can* escape the crisis. We can return to the community of life, we can re-situate ourselves, in thought and experience, *within* and not *against* the more-than-human world. But we will not do so if we continue to suppose that we face only a practical problem: how to recycle better or pollute less or save the rainforests. Yes, yes, all of this is crucial too. The fundamental suggestion in this book, though, and truly the working hypothesis of environmental philosophy, is that the environmental crisis ultimately lies deeper, challenges us more profoundly, more *philosophically*, and offers us unsuspected and fabulous opportunities too. This is what we invite you to explore. Welcome to the adventure.

III

Next a word about the contributors—but really a word about the keys in which "philosophy" will be offered here.

Each of us, I think, have come to (are *compelled to*) environmental philosophy out of a sense both of love and pain. Certain places call our names: the Australian bush; the Eastern Front of the Rockies; for me, Wisconsin's Sand Counties. I can never hear a whippoorwill—the clear-voiced, night-singing phantom bird of that land—without crying: crying from sheer joy for coming into life roaming that land, and for loss, now, as the whippoorwills are distant and rare, lost to pesticides and the clearing of the copses. This is *me*: it is that little ground-dweller, pestbird of the night (the neighbors back on the prairie used to fire off shotguns to scare them off so we could sleep), that calls my name. Now, city dweller half a continent away, I dream of waking to their calls, a dream of youth, like a dream of first love. Meadowlarks have mostly left the fields around home too, my mother's bird, lilting morning sprite.

And it is out of this that our philosophy comes. The hills and the birds called our names long before we were captivated by this rearrangement of abstractions we call philosophy. But philosophy now opens one way for us to give them voice.

My own essay in this volume speaks of exactly this: both of the staggering losses the biosphere is suffering, and of the possibilities that nonetheless remain, sometimes in the most hidden and unexpected places. The losses are heartbreaking. The possibilities are, not surprisingly, less easy to see, so here the writing does indeed turn speculative. Imagine the world as it could be once again. It is possible to begin to rethink the ways the land is co-inhabited, for example to restore the little patches of inviting habitat that would call back the birds, and to recover a sense of the life that is still and must be here. To think this way is already to imagine a new transformative practice: to see more deeply and completely into the world we think we know, and to reclaim and reinvoke the hidden possibilities of things—and so, already, an essentially philosophical project too. It is to go, once again, *beyond* crisis to a vision of the world as it *could* be.

Likewise, David Abram, reaching for a way to speak of the obscured but wild possibilities of the senses, mixes the story of his travels in Indonesia with speculations in cultural history and a theory of the shaman's magic. This too is philosophical. Yet it remains profoundly *grounded*: indeed, quite palpably grounded. We speak from certain places, even of the soil and its denizens, in the company of a larger kind of intelligence that Abram calls "more-than-human." This is what he discovered—this is what grew upon him—in Indonesia, across the years in New Mexico, and now in his new home, the Pacific Northwest:

> Oak, madrone, Douglas fir, red-tailed hawk, serpentine in the sandstone, a certain scale to the topography, drenching rains in the winter, fog off-shore in the summer, salmon surging in the streams—all of these together make up a particular state of mind, a place-specific intelligence shared by all the humans that dwell therein, but also by the coyotes yapping in those valleys, by the bobcats and the ferns and the spiders, by all beings who live and make their way in that zone. Each place its own psyche. Each sky its own blue.

Philosophy for Abram becomes the most thorough-going of ways to bring all of this back into the realm of direct sensory experience, to open the cultural space for it once again. What would it be, I asked earlier, to rejoin the community of life? Here we begin to sense—to quite literally *sense*—an answer.

Val Plumwood writes of her childhood in the Australian bush:

> I was tutored at home, and formally registered, like so many bush children of the time, in Correspondence School. But the bush was my real school, and supplied most of my friends, adventures and conversations, often inspired by my favorite fictional character, the fearless and philosophical Alice. . . .
>
> Even after a brushfire swept through our forest, disclosing unguessed-at neighbors like the lively frilled lizard the fireman held up for us children to gasp at, the forest seemed inexhaustible. Soon the blackened forest was in leaf again, and it still brought forth the August gems, the purple glossodia (waxlip) orchids I liked to gather for my birthday flower. Don't pick too many, my wise mother warned, or you'll lose them. But I did, and the numbers in my favorite patch on the ridge declined, until one year they appeared no more.
>
> That was the beginning for me of two things: the first, a long love-affair with the color purple. . . . And also the first glimpse, a speck entering the edge of my field of vision, of something that now looms large over the landscape of life. This is the knowledge that the magical and apparently invincible world of

the forest is fragile and vulnerable, that its loss could be imminent, permanent, and irreversible: that the great, intricate, unfathomable forest of childhood, the apparently immortal trav-
elor who has crossed vast ages of planetary time, may already be
critically injured, and could go from the earth. . . . Now [it] *depends upon our care and our will to fight.* . . .

Pain and loss again, and so one may be tempted to read this
kind of account as only "nostalgic," sentimental. We yearn for lost
childhood, innocence, an unthreatened world. That is nothing to
apologize for: why *shouldn't* we yearn for an unthreatened world?
But as it becomes philosophical, as it comes to stand behind philosophical work, it also takes on a more powerful hue. Like the more-
than-human everywhere, the "school of the bush" offers a sense of
what the world *could* be like, what a human or humane relation to
nature could be, even right now in this world, the techno-world at
the beginning of a new Millennium. We still *are* animals and beings of nature, part of this Earth, and so when Plumwood argues
that species-egoism is a philosophical illusion, she is speaking from
real experience. As simple as that. From the "school of the bush"
one can't encounter the species-egoist claim without a certain
amazement. It is somehow to mislay a whole dimension of existence. And this is what Plumwood, herself "fearless and philosophical" like her heroine Alice, lets us hear.

Holmes Rolston, avid naturalist, is a guide extraordinaire, both
of the true wilderness and what he calls "the deep wilderness of values." Are they not more than a little similar? "For the trip you are
about to take I offer myself as a wilderness guide. . . . A century ago
the challenge was to know where you were geographically in a blank
spot on the map, but today we are bewildered philosophically in
what has long been mapped as a moral blank space. . . . Values run
off our maps. . . ." So philosophy becomes another form of wilderness exploration, and are we not to think that the scientific naturalist's attentiveness and passion apply to it as well? The territory may
be more theoretical—the new regions through which Rolston guides
us here are the emerging levels of environmental ethics, regions that

may invite new ethical frameworks and theories—but the trip unfolds flowingly, each step beyond the familiar is well prepared and at every turn shows the naturalist's attention to detail. At these heights of abstraction too we may learn to feel at home.

Environmental philosophy—the academic version—began with the promise, as Jim Cheney puts it, "of a new ethic, one which would go beyond traditional Euro-American, human-centered ethics to include . . . some, perhaps all, of non-human nature. . . ." He notes that things have mostly not turned out that way. Mostly we still mind the boundary markers and warning signs erected by our training and culture: the signs that tell us that only humans and the human-like count; not to mind the rocks; and that even the animals have to prove themselves if they are to "count" at all. But Cheney is an explorer, bushwhacker, naturalist too, and free spirit. Like most back-country types, he does not have much time for warning signs and boundary markers and following the familiar old paths. He is more likely to shoulder a backpack and vanish in a second if not watched carefully. So Cheney closes this book, fittingly, with a philosophical version of the same thing: a hint of the paths as yet unexplored in our official philosophies; a vanishing into the landscape of place-narrative, of native peoples' practice (insofar as we of European heritage can enter that world: not as far as we might like to think), of storied rocks, Tricksters, ceremonial prairie burns, the substance of his own life.

So once again: what we write here, our philosophy itself, flows from our lives, from what we have been privileged to know of the wild and from the wild in ourselves. These five distinctive philosophical voices are also five distinctive personal voices. We speak from our own losses, from our struggles for the sense of possibility, to reclaim the depths of the world as it is and to reinvoke its hidden possibilities. Of course we also have our differences, even quite fundamental ones, though not much is made of them here. This book is meant as an invitation, an opening of a field of possibility, not some kind of debate. Look, first and foremost, for what we share. Look for the life and the spirit within—beside, behind, beyond—the words.

IV

This book is meant as a contribution to environmental philosophy as well as a primer in it; and environmental philosophy itself represents a contribution to philosophy as a whole, not just philosophy's contribution to environmental thinking. All of this takes some explaining.

Academic philosophy in the English-speaking world passed through a period around mid-century when (at least in the views of many of us) a narrow and technical view of the field took hold. Philosophers were supposed to be specialists in certain kinds of argument and conceptual analysis, clarifying the claims of others and cutting them down to size. The grand unities aspired to by nineteenth-century philosophers were supposed to be unattainable and confused; the social reconstruction practiced by early twentieth-century pragmatists was supposed to be unprofessional and politically compromising.

Since the 1970s, however, philosophy has dramatically reemerged from this self-spun cocoon. Large social issues are again at the forefront, though philosophers may approach them with a distinctive set of conceptual tools. Continental philosophers— French, German, Dutch—are once again attended to, with their phenomenological and now "postmodern" preoccupations increasingly shared by English and American philosophers. And for both of these reasons in turn, *direct experience*—an appreciation for the richness of the world and the depths at which we are enmeshed in it and indeed constituted by it—is once again central to philosophical attention as well.

Environmental philosophy is part of this development, and its turn may even have come to nudge the same development farther. At the very moment when philosophers began to attend again to the richness and depth of the human world—from our physical embodiment to the multiple levels at which we are literally made of words, or made of culture—environmental philosophy emerged to insist upon the next step: recognizing the richness and depth of the *more*-than-human world: the way in which physical embodiment, for example, does not end at the skin (the so-called "skin-encapsulated ego" was an early target of ecological thinkers), and the ways in which we are and must always be profoundly *animal*— a distinctive animal, perhaps, but so are *all* animals—as well as kin even to stone and stars. Very large issues come up, from the cultural history of our alienation from nature to the resources for recovering a sense of connection *now*.

Yet the older and more professionalized aspirations for philosophy also live on. The character of philosophy itself is an essentially contested question, one perhaps never to be settled but only debated over by shifting opponents. Moreover, older and more cautious conceptions always have a good deal of staying power, mainly because professional philosophers are almost all academics, for whom not only caution and precision but also specialization and certain conventions of "voice" are deep norms, difficult to change or resist.

The result is that (to oversimplify *a bit*) there is a pronounced division in environmental philosophy. On the one hand, there is a body of writing that is fairly professional; objective, theoretical, and cautious in voice; and academic rather than political or spiritual in its concerns. It is also mostly "extensionist" in its approach: that is, the tendency is to work outward from well-understood (which is to say, traditionally human-centered) ethical paradigms to try to accomodate environmental concerns. Sometimes the result is a kind of enlightened humanism; sometimes a self-conscious break with humanism, but where the new ethic, all the same, has most of the same elements and even implications as the old ones. English and American academics dominate this field, though professional courtesy is extended to the few available sympathetically presented Continental approaches.

On the other side, there is a much wilder, more speculative and experience-based kind of writing, usually much closer to particular places or struggles; more concerned with action; often narrative, mythical, or poetic. This literature is resolutely nonextensionist: that is, the going view is that the environmental crisis requires of us a completely new way of thinking of ourselves and the earth. Tinkering with the old ethics will not do it. Philosophy in this key has a toe-hold in the academy, but no more.

Most of the writing is by activists, organizers, poets, and nature writers. What professional philosophers do show up are usually Continental thinkers or more theological figures or writers from academically more marginalized places, like bioregionalism and ecofeminism.

This polarization is both unfortunate and unnecessary. Both sides lose on account of it. Too often the academic style enforces a kind of abstraction and endless hedging in exactly the places where we want most to know what to *do*, and whether or not it is "too late." Also, the necessarily detached voice often pulls head and heart apart. On the other hand, the activist/poetic tradition too often lacks critical care; the claims are too easy, or too sweeping or unclear. People also want to know how to *think*, facing in some ways an unprecedented philosophical situation, and now, after the first wave of vertigo and amazement has passed, we recognize that this kind of thinking is *hard*. As Rolston says, values are running off the maps. In real-world wilds, those who follow blindly sometimes don't come back.

The upshot is a truism, but seldom heeded for all that: each side needs the other's help. We need to think more wildly and widely, and *also* more critically and carefully. We need to think globally *and* locally, and act globally *and* locally. We need to take heed of scientific and Continental and Eastern and indigenous voices, but without just appropriating them: we need to learn from them with due care or respect—which is not as easy as it may seem—and we need to maintain some critical distance too. We need to answer the questions that people across the culture are rightly asking, not just (but also including) the questions asked by academics, activists, and other kinds of hermits.

I would like to think that in a small way this book is a contribution to putting these two sides, these two strands, back together: toward answering the real questions. All of us who have contributed to this book feel caught, I think, to some extent in between. We have each contributed in our ways to both camps, to both bodies of thinking and writing. We have all tried in our best work to bring them together—and that is the kind of work I have tried to bring together, in turn, in this book.

Readers will have to judge for themselves how far we have succeeded. This is, in any case, our aim: to think in a way that is concrete and yet wide-angled all at once. To give political and other kinds of action a philosophical context, at the same time that the debate over the airiest abstractions finds itself drawn back to real, on-the-ground struggles. To follow the values that "run off the maps," but at a safe enough distance to be able to keep our bearings as we go. And to return "home"—a concept crucial to both Rolston's essay and Cheney's—while understanding at the same time that "home" too is a place that we still need to get to know. We will find as necessary guides everyone from the crickets and the octopus to our own Zen poets, ecologists and biologists, native elders, and, yes, even professional philosophers.

V

Finally, a few words on the uses of this book.

This book should be accessible to all interested readers, whether or not they are taking a class in environmental philosophy or related subjects. We have simply tried to offer a short, accessible, provocative work that motivates the central questions, tells compelling stories, and offers a range of philosophical responses, not as antagonists but as complementary explorations: a true *invitation* to environmental philosophy. We have taken care to keep the language accessible and the academic baggage to a minimum.

Each of the essays represents environmental philosophy in a somewhat different style. No essay exactly "surveys" the field. None of us campaign for a certain philosophical method or style, or are even very explicit about such questions. We have aimed simply to exemplify different styles at their best, offering complementary ways of engaging real questions and opening up real possibilities.

Each essay also opens doors into larger literatures and debates—philosophical, ecological, and others—and so the epilogue, "Going On," offers a bibliographical guide to each of these areas, as well as concrete and practical "next steps." The epilogue is also meant for all readers, academic or not. Suggestions for further reading go hand in hand with the names and addresses of environmental organizations, ideas for celebrating new holidays, and the like.

For class use, this book might be assigned as a first reading in an environmental ethics or environmental philosophy course, or its essays might be used separately as introductions to environmental philosophy in different styles. As a first reading, this book might be followed up in a variety of ways: with a standard anthology in environmental ethics and/or with other classic or provocative works in the field. Many possible further readings are cited in the epilogue. A variety of exercises and applications are also possible: again, see the epilogue for some suggestions and resources.

As they are arranged here, each piece opens the way for the next, although they can also stand alone or in some other order. The essays might also be used separately or in pairs to open up some range of questions and some part of the larger literature. For example, Abram's piece and my own might be used to open a course, bringing into focus some basic underlying questions about the more-than-human world and our relation to it. What *are* the powers of the earth, and of other creatures? How "open" is our situation: what are the hidden possibilities—hidden *invitations*, we might now say, bearing in mind this book's title—in the present situation, that crisis that in some ways we think we know so well?

Plumwood's and Rolston's essays could be used to introduce (to invite students into) the traditional environmental ethics literature in the anthologies. Plumwood analyzes one of the essential concepts in that literature; Rolston surveys some of the theoretical positions in environmental ethics. Students then have an overview they can carry into more detailed readings.

Finally, Cheney's essay might be used to introduce a closing section of a course, with attention to quite different kinds of voices, specifically Native American in his essay, and to the question of ritual, the great cycles of "ceremonial renewal and regeneration." An invitation to practice, and back to the wider world, at the end. But of course this is only one of many possible arrangements.

It is our hope that this book will prove inviting not only to interested students and newcomers to the field—a way into the field that compels and intrigues—but also at the same time that it will prove a little refreshing and perhaps even provocative (perhaps more than a *little* provocative) to those already involved with the field. In this way the very division between two strands in environmental philosophy just discussed becomes an opportunity, itself a kind of invitation. A book that puts the two sides back together again ought to be challenging and inviting for readers at all levels, not least for philosophers looking for new and more inclusive ways to approach and practice environmental philosophy itself.

I have spoken of all the contributors' love of the land, of our varieties of wild experience. It is also just as true that we all love philosophy, and it is out of this passion too that we write. It is our hope that *both* passions will grow on our readers. So, once again, welcome to the adventure. Though this may be your first trip to this particular wild country, we hope—as with most trips to wild country, real *or* philosophical—that it will be but the first of many.

A. W.

1

A More-Than-Human World

David Abram

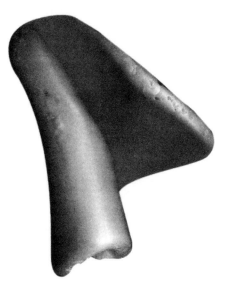

Late one evening I stepped out of my little hut in the rice paddies of eastern Bali and found myself falling through space. Over my head the black sky was rippling with stars, densely clustered in some regions, almost blocking out the darkness between them, and more loosely scattered in other areas, pulsing and beckoning to each other. Behind them all streamed the great river of light with its several tributaries. Yet the Milky Way churned beneath me as well, for my hut was set in the middle of a large patchwork of rice paddies, separated from each other by narrow two-foot-high dikes, and these paddies were all filled with water. The surface of these pools, by day, reflected perfectly the blue sky, a reflection broken only by the thin, bright green tips of new rice. But by night the stars themselves glimmered from the surface of the paddies, and the river of light whirled through the darkness underfoot as well as above; there seemed no ground in front of my feet, only the abyss of star-studded space falling away forever.

I was no longer simply beneath the night sky, but also *above* it—the immediate impression was of weightlessness. I might have been able to reorient myself, to regain some sense of ground and gravity, were it not for a fact that confounded my senses entirely: between the constellations below and the constellations above drifted countless fireflies, their lights flickering like the stars, some drifting up to join the clusters of stars overhead, others, like graceful meteors, slipping down from above to join the constellations underfoot, and all these paths of light upward and downward were mirrored, as well, in the still surface of the paddies. I felt myself at times falling through space, at other moments floating and drifting. I simply could not dispel the profound vertigo and giddiness; the paths of the fireflies, and their reflections in the water's surface, held me in a sustained trance. Even after I crawled back to my hut and shut the door on this whirling world, I felt that now the little room in which I lay was itself floating free of the earth.

Fireflies! It was in Indonesia, you see, that I was first introduced to the world of insects, and there that I first learned of the great influence that insects—such diminutive entities—could have upon the human senses. I had traveled to Indonesia on a research grant to study magic—more precisely, to study the relation between magic and medicine, first among the traditional sorcerers, or *dukuns*, of the Indonesian archipelago, and later among the *dzankris*, the traditional shamans of Nepal. One aspect of the grant was somewhat unique: I was to journey into rural Asia not outwardly as an anthropologist or academic researcher, but as a magician in

my own right, in hopes of gaining a more direct access to the local sorcerers. I had been a professional sleight-of-hand magician for
five years back in the United States, helping to put myself through
college by performing in clubs and restaurants throughout New
England. I had, as well, taken a year off from my studies in the
psychology of perception to travel as a street magician through Europe and, toward the end of that journey, had spent some months
in London, England, exploring the use of sleight-of-hand magic in
psychotherapy, as a means of engendering communication with
distressed individuals largely unapproachable by clinical healers.
The success of this work suggested to me that sleight-of-hand
might lend itself well to the curative arts, and I became, for the
first time, interested in the relation, largely forgotten in the West,
between folk medicine and magic.

It was this interest that led to the aforementioned grant, and to
my sojourn as a magician in rural Asia. There, my sleight-of-hand
skills proved invaluable as a means of stirring the curiosity of the
local shamans. For magicians—whether modern entertainers or indigenous, tribal sorcerers—have in common the fact that they
work with the malleable texture of perception. When the local sorcerers gleaned that I had at least some rudimentary skill in altering the common field of perception, I was invited into their homes,
asked to share secrets with them, and eventually encouraged, even
urged, to participate in various rituals and ceremonies.

But the focus of my research gradually shifted from questions
regarding the application of magical techniques in medicine and
ritual curing toward a deeper pondering of the relation between
traditional magic and the animate natural world. This broader concern seemed to hold the keys to the earlier questions. For none of
the several island sorcerers that I came to know in Indonesia, nor
any of the *dzankris* with whom I lived in Nepal, considered their
work as ritual healers to be their major role or function within
their communities. Most of them, to be sure, *were* the primary
healers or "doctors" for the villages in their vicinity, and they were
often spoken of as such by the inhabitants of those villages. But
the villagers also sometimes spoke of them, in low voices and in
very private conversations, as witches (or "lejaks" in Bali), as dark

magicians who at night might well be practicing their healing spells backward (or while turning to the left instead of to the right) in order to afflict people with the very diseases that they would later work to cure by day. Such suspicions seemed fairly common in Indonesia, and often were harbored with regard to the most effective and powerful healers, those who were most renowned for their skill in driving out illness. For it was assumed that a magician, in order to expel malevolent influences, must have a strong understanding of those influences and demons—even, in some areas, a close rapport with such powers. I myself never consciously saw any of those magicians or shamans with whom I became acquainted engage in magic for harmful purposes, nor any convincing evidence that they had ever done so. (Few of the magicians that I came to know even accepted money in return for their services, although they did accept gifts in the way of food, blankets, and the like.) Yet I was struck by the fact that none of them ever did or said anything to counter such disturbing rumors and speculations, which circulated quietly through the regions where they lived. Slowly, I came to recognize that it was through the agency of such rumors, and the ambiguous fears that such rumors engendered in the village people, that the sorcerers were able to maintain a basic level of privacy. If the villagers did not entertain certain fears about the local sorcerer, then they would likely come to obtain his or her magical help for every little malady and disturbance; and since a more potent practitioner must provide services for several large villages, the sorcerer would be swamped from morning to night with requests for ritual aid. By allowing the inevitable suspicions and fears to circulate unhindered in the region (and sometimes even encouraging and contributing to such rumors), the sorcerer ensured that *only* those who were in real and profound need of his skills would dare to approach him for help.

This privacy, in turn, left the magician free to attend to what he acknowledged to be his primary craft and function. A clue to this function may be found in the circumstance that such magicians rarely dwell at the heart of their village; rather, their dwellings are commonly at the spatial periphery of the community or, more often, out beyond the edges of the village—amid the rice fields, or in a forest, or a wild cluster of boulders. I could easily attribute this to the just-mentioned need for privacy, yet for the magician in a traditional culture it seems to serve another purpose as well, providing a spatial expression of his or her symbolic position with regard to the community. For the magician's intelligence is not encompassed *within* the society; its place is at the edge of the community, mediating *between* the human community and the larger community of beings upon which the village depends for its nourishment and sustenance. This larger community includes, along with the humans, the multiple nonhuman entities that constitute the local landscape, from the diverse plants and the myriad animals—birds, mammals, fish, reptiles, insects—that inhabit or migrate through the region, to the particular winds and weather patterns that inform the local geography, as well as the various landforms—forests, rivers, caves, mountains—that lend their specific character to the surrounding earth.

The traditional or tribal shaman, I came to discern, acts as an intermediary between the human community and the larger ecological field, ensuring that there is an appropriate flow of nourishment, not just from the landscape to the human inhabitants, but from the human community back to the local earth. By his constant rituals, trances, ecstasies, and "journeys," he ensures that the relation between human society and the larger society of beings is balanced and reciprocal, and that the village never takes more from the living land than it returns to it—not just materially but with prayers, propitiations, and praise. The scale of a harvest or the size of a hunt are always negotiated between the tribal community and the natural world that it inhabits. To some extent every adult in the community is engaged in this process of listening and attuning to the other presences that surround and influence daily life. But the shaman or sorcerer is the exemplary voyager in the intermediate realm between the human and the more-than-human worlds, the primary strategist and negotiator in any dealings with the Others.

And it is only as a result of her continual engagement with the animate powers that dwell beyond the human community that the traditional magician is able to alleviate many individual illnesses that arise *within* that community. The sorcerer derives her ability to cure ailments from her more continuous practice of "healing" or balancing the community's relation to the surrounding land. Disease, in such cultures, is often conceptualized as a kind of systemic imbalance within the sick person, or more vividly as the intrusion of a demonic or malevolent presence into his body. There are, at times, malevolent influences within the village or tribe itself that disrupt the health and emotional well-being of susceptible individuals within the community. Yet such destructive influences within the human community are commonly traceable to a disequilibrium between that community and the larger field of forces in which it is embedded. Only those persons who, by their everyday practice, are involved in monitoring and maintaining the relations *between* the human village and the animate landscape are able to appropriately diagnose, treat, and ultimately relieve personal ailments and illnesses arising *within* the village. Any healer who was not simultaneously attending to the intertwined relation between the human community and the larger, more-than-human field, would likely dispel an illness from one person only to have the same problem arise (perhaps in a new guise) somewhere else in the community. Hence, the traditional magician or medicine person functions primarily as an intermediary between human and nonhuman worlds, and only secondarily as a healer. Without a continually adjusted awareness of the relative balance or imbalance between the human group and its nonhuman environ, along with the skills necessary to modulate that primary relation, any "healer" is worthless—indeed, not a healer at all. The medicine person's primary allegiance, then, is not to the human community, but to the earthly web of relations in which that community is embedded—it is from this that his or her power to alleviate human illness derives—and this sets the local magician apart from other persons.

The primacy for the magician of nonhuman nature—the centrality of his relation to other species and to the earth—is not always evident to Western researchers. Countless anthropologists have managed to overlook the ecological dimension of the shaman's craft, while writing at great length of the shaman's rapport with

"supernatural" entities. We can attribute much of this oversight to
the modern, civilized assumption that the natural world is largely
determinate and mechanical, and that that which is regarded as
mysterious, powerful, and beyond human ken must therefore be of
some other, nonphysical realm *above* nature, "supernatural."

The oversight becomes still more comprehensible when we realize that many of the earliest European interpreters of indigenous
lifeways were Christian missionaries. For the Church had long assumed that only human beings have intelligent souls, and that the
other animals, to say nothing of trees and rivers, were "created" for
no other reason than to serve humankind. We can easily understand why European missionaries, steeped in the dogma of institutionalized Christianity, assumed a belief in supernatural, other-
worldly powers among those tribal persons whom they saw
awestruck and entranced by nonhuman (but nevertheless natural)
forces. What is remarkable is the extent to which contemporary
anthropology still preserves the ethnocentric bias of these early interpreters. We no longer describe the shamans' enigmatic spirit-
helpers as the "superstitious claptrap of heathen primitives"—we
have cleansed ourselves of at least *that* much ethnocentrism; yet we
still refer to such enigmatic forces, respectfully now, as "supernatural"—for we are unable to shed the sense, so endemic to scientific
civilization, of nature as a rather prosaic and predictable realm, un-
suited to such mysteries. Nevertheless, that which is regarded with
the greatest awe and wonder by indigenous, oral cultures is, I suggest, none other than what we view as nature itself. The deeply
mysterious powers and entities with whom the shaman enters into
a rapport are ultimately the same forces—the same plants, animals,
forests, and winds—that to literate, "civilized" Europeans are just
so much scenery, the pleasant backdrop of our more pressing human concerns.

The most sophisticated definition of "magic" that now circulates through the American counterculture is "the ability or power
to alter one's consciousness at will." No mention is made of any
reason for altering one's consciousness. Yet in tribal cultures that
which we call "magic" takes its meaning from the fact that humans, in an indigenous and oral context, experience their own con-

sciousness as simply one form of awareness among many others. The traditional magician cultivates an ability to shift out of his or her common state of consciousness precisely in order to make contact with the other organic forms of sensitivity and awareness with which human existence is entwined. Only by temporarily shedding the accepted perceptual logic of his culture can the sorcerer hope to enter into relation with other species on their own terms; only by altering the common organization of his senses will he be able to enter into a rapport with the multiple nonhuman sensibilities that animate the local landscape. It is this, we might say, that defines a shaman: the ability to readily slip out of the perceptual boundaries that demarcate his or her particular culture—boundaries reinforced by social customs, taboos, and most importantly, the common speech or language—in order to make contact with, and learn from, the other powers in the land. His magic is precisely this heightened receptivity to the meaningful solicitations—songs, cries, gestures—of the larger, more-than-human field.

Magic, then, in its perhaps most primordial sense, is the experience of existing in a world made up of multiple intelligences, the intuition that every form one perceives—from the swallow swooping overhead to the fly on a blade of grass, and indeed the blade of grass itself—is an *experiencing* form, an entity with its own predilections and sensations, albeit sensations that are very different from our own.

To be sure, the shaman's ecological function, his or her role as intermediary between human society and the land, is not always obvious at first blush, even to a sensitive observer. We see the sorcerer being called upon to cure an ailing tribesman of his sleeplessness, or perhaps simply to locate some missing goods; we witness him entering into trance and sending his awareness into other dimensions in search of insight and aid. Yet we should not be so ready to interpret these dimensions as "supernatural," nor to view them as realms entirely "internal" to the personal psyche of the practitioner. For it is likely that the "inner world" of our Western psychological experience, like the supernatural heaven of Christian belief, originates in the loss of our ancestral reciprocity with the animate earth. When the animate powers that surround us are suddenly construed as having less significance than ourselves, when the generative earth is abruptly defined as a determinate object devoid of its own sensations and feelings, then the sense of a wild and multiplicitous otherness (in relation to which human existence has always oriented itself) must migrate, either into a supersensory heaven beyond the natural world, or else into the human skull itself—the only allowable refuge, in this world, for what is ineffable and unfathomable.

But in genuinely oral, indigenous cultures, the sensuous world itself remains the dwelling place of the gods, of the numinous powers that can either sustain or extinguish human life. It is not by sending his awareness out beyond the natural world that the shaman makes contact with the purveyors of life and health, nor by journeying into his personal psyche; rather, it is by propelling his awareness laterally, outward into the depths of a landscape at once both sensuous and psychological, the living dream that we share with the soaring hawk, the spider, and the stone silently sprouting lichens on its coarse surface.

The magician's intimate relationship with nonhuman nature becomes most evident when we attend to the easily overlooked background of his or her practice—not just to the more visible tasks of curing and ritual aid to which she is called by individual clients, or to the larger ceremonies at which she presides and dances, but to the content of the prayers by which she prepares for such ceremonies, and to the countless ritual gestures that she enacts when alone, the daily propitiations and praise that flow from her toward the land and *its* many voices.

All this attention to nonhuman nature was, as I have mentioned, very far from my intended focus when I embarked on my research into the uses of magic and medicine in Indonesia, and it was only gradually that I became aware of this more subtle dimension of the native magician's craft. The first shift in my preconceptions came rather quietly, when I was staying for some days in the home of a young "balian," or magic practitioner, in the interior of Bali. I had been provided with a simple bed in a separate, one-room building in the balian's family compound (most compound homes, in Bali, are comprised of several separate small buildings, for sleeping and for cooking, set on a single enclosed plot of land), and early each morning the balian's wife came to bring me a small but delicious bowl of fruit, which I ate by myself, sitting on the ground outside, leaning against the wall of my hut and watching the sun slowly climb through the rustling palm leaves. I noticed, when she delivered the fruit, that my hostess was also balancing a tray containing many little green plates: actually, they were little boat-shaped platters, each woven simply and neatly from a freshly cut section of palm frond. The platters were two or three inches long, and within each was a little mound of white rice. After handing me my breakfast, the woman and the tray disappeared from view behind the other buildings, and when she came by some minutes later to pick up my empty bowl, the tray in her hands was empty as well.

The second time that I saw the array of tiny rice platters, I asked my hostess what they were for. Patiently, she explained to me that they were offerings for the household spirits. When I inquired about the Balinese term that she used for "spirit," she repeated the same explanation, now in Indonesian, that these were gifts for the spirits of the family compound, and I saw that I had understood her correctly. She handed me a bowl of sliced papaya and mango, and disappeared around the corner. I pondered for a minute, then set down the bowl, stepped to the side of my hut, and peered through the trees. At first unable to see her, I soon caught sight of her crouched low beside the corner of one of the other buildings, carefully setting what I presumed was one of the offerings on the ground at that spot. Then she stood up with the tray, walked to the other visible corner of the same building, and there slowly and carefully set another offering on the ground. I returned to my bowl of fruit and finished my breakfast. That afternoon, when the rest of the household was busy, I walked back behind the building where I had seen her set down the two offerings. There were the little green platters, resting neatly at the two rear corners of the building. But the mounds of rice that had been within them were gone.

The next morning I finished the sliced fruit, waited for my hostess to come by for the empty bowl, then quietly headed back behind the buildings. Two fresh palm-leaf offerings sat at the same spots where the others had been the day before. These were filled with rice. Yet as I gazed at one of these offerings, I abruptly realized, with a start, that one of the rice kernels was actually moving. Only when I knelt down to look more closely did I notice a line of tiny black ants winding through the dirt to the offering. Peering still closer, I saw that two ants had already climbed onto the offering and were struggling with the uppermost kernel of rice; as I watched, one of them dragged the kernel down and off the leaf, then set off with it back along the line of ants advancing on the offering. The second ant took another kernel and climbed down with it, dragging and pushing, and fell over the edge of the leaf, then a third climbed onto the offering. The line of ants seemed to emerge from a thick clump of grass around a nearby palm tree. I walked over to the other offering and discovered another line of ants dragging away the white kernels. This line emerged from the top of a little mound of dirt, about fifteen feet away from the buildings. There was an offering on the ground by a corner of my building as well, and a nearly identical line of ants. I walked into my room chuckling to myself: the balian and his wife had gone to so much trouble to placate the household spirits with gifts, only to have their offerings stolen by little six-legged thieves. What a waste! But then a strange thought dawned on me: what if the ants were the very "household spirits" to whom the offerings were being made?

I soon began to discern the logic of this. The family compound, like most on this tropical island, had been constructed in the vicinity of several ant colonies. Since a great deal of cooking took place in the compound (which housed, along with the balian and his wife and children, various members of their extended family), and also much preparation of elaborate offerings of foodstuffs for various rituals and festivals in the surrounding villages, the grounds and the buildings at the compound were vulnerable to infestations by the sizable ant population. Such invasions could range from rare nuisances to a periodic or even constant siege. It became apparent that the daily palm-frond offerings served to preclude such an attack by the natural forces that surrounded (and underlay) the family's land. The daily gifts of rice kept the ant colonies occupied—and, presumably, satisfied. Placed in regular, repeated locations at the corners of various structures around the compound, the offerings seemed to establish certain boundaries between the human and ant communities; by honoring this boundary with gifts, the humans apparently hoped to persuade the insects to respect the boundary and not enter the buildings.

Yet I remained puzzled by my hostess's assertion that these were gifts "for the spirits." To be sure, there has always been some confusion between our Western notion of "spirit" (which so often is defined in contrast to matter or "flesh"), and the mysterious presences to which tribal and indigenous cultures pay so much respect. I have already alluded to the gross misunderstandings arising from the circumstance that many of the earliest Western students of these other customs were Christian missionaries all too ready to see occult ghosts and immaterial phantoms where the tribespeople were simply offering their respect to the local winds. While the notion of "spirit" has come to have, for us in the West, a primarily anthropomorphic or human association, my encounter with the ants was the first of many experiences suggesting to me that the "spirits" of an indigenous culture are primarily those modes of intelligence or awareness that do *not* possess a human form.

As humans, we are well acquainted with the needs and capacities of the human body—we *live* our own bodies and so know, from within, the possibilities of our form. We cannot know, with the same familiarity and intimacy, the lived experience of a grass snake or a snapping turtle; we cannot readily experience the precise sensations of a hummingbird sipping nectar from a flower or a rubber tree soaking up sunlight. And yet we do know how it feels to sip from a fresh pool of water or to bask and stretch in the sun. Our experience may indeed be a variant of these other modes of sensitivity; nevertheless, we cannot, as humans, precisely experience the living sensations of another form. We do not know, with full clarity, their desires or motivations; we cannot know, or can never be sure that we know, what they know. That the deer does experience sensations, that it carries knowledge of how to orient in the land, of where to find food and how to protect its young, that it knows well how to survive in the forest without the tools upon which we depend, is readily evident to our human senses. That the mango tree has the ability to create fruit, or the yarrow plant the power to reduce a child's fever, is also evident. To humankind, these Others are purveyors of secrets, carriers of intelligence that we ourselves often need: it is these Others who can inform us of unseasonable changes in the weather, or warn us of imminent eruptions and earthquakes, who show us, when foraging, where we may find the ripest berries or the best route to follow back home. By watching them build their nests and shelters, we glean clues regarding how to strengthen our own dwellings, and their deaths teach us of our own. We receive from them countless gifts of food, fuel, shelter, and clothing. Yet still they remain Others to us, inhabiting their own cultures and displaying their own rituals, never wholly fathomable.

Moreover, it is not only those entities acknowledged by Western civilization as "alive," not only the other animals and the plants that speak, as spirits, to the senses of an oral culture, but also the meandering river from which those animals drink, and the torrential monsoon rains, and the stone that fits neatly into the palm of the hand. The mountain, too, has its thoughts. The forest birds whirring and chattering as the sun slips below the horizon are vocal organs of the rain forest itself.

Bali, of course, is hardly an aboriginal culture; the complexity of its temple architecture, the intricacy of its irrigation systems, the resplendence of its colorful festivals and crafts all bespeak the influence of various civilizations, most notably the Hindu complex of India. In Bali, nevertheless, these influences are thoroughly intertwined with the indigenous animism of the Indonesian archipelago; the Hindu gods and goddesses have been appropriated, as it were, by the more volcanic, eruptive spirits of the local terrain.

Yet the underlying animistic cultures of Indonesia, like those of many islands in the Pacific, are steeped as well in beliefs often referred to by ethnologists as "ancestor worship," and some may argue that the ritual reverence paid to one's long-dead human ancestors (and the assumption of their influence in present life), easily invalidates my assertion that the various "powers" or "spirits" that move through the discourse of indigenous, oral peoples are ultimately tied to nonhuman (but nonetheless sentient) forces in the enveloping landscape.

This objection rests upon certain assumptions implicit in Christian civilization, such as the assumption that the "spirits" of dead persons necessarily retain their human form, and that they reside in a domain outside of the physical world to which our senses give us access. However, most indigenous tribal peoples have no such ready recourse to an immaterial realm outside earthly nature. Our strictly human heavens and hells have only recently been abstracted from the sensuous world that surrounds us, from this more-than-human realm that abounds in its own winged intelligences and clovenhoofed powers. For almost all oral cultures, the enveloping and sensuous earth remains the dwelling place of both the living *and* the dead. The "body"—whether human or otherwise—is not yet a mechanical object in such cultures, but is a magical entity, the mind's own sensuous aspect, and at death the body's decomposition into soil, worms, and dust can only signify the gradual reintegration of one's ancestors and elders into the living landscape, from which all, too, are born.

Each indigenous culture elaborates this recognition of metamorphosis in its own fashion, taking its clues from the particular terrain in which it is situated. Often the invisible atmosphere that animates the visible world—the subtle presence that circulates both within us and between all things—retains within itself the spirit or breath of the dead person until the time when that breath will enter and animate another visible body—a bird, or a deer, or a field of wild grain. Some cultures may burn, or "cremate," the body in order to more completely return the person, as smoke, to the swirling air, while that which departs as flame is offered to the sun and stars, and that which lingers as ash is fed to the dense earth. Still other cultures may dismember the body, leaving certain parts in precise locations where they will likely be found by condors, or where they will be consumed by mountain lions or by wolves, thus hastening the re-incarnation of that person into a particular animal realm within the landscape. Such examples illustrate simply that death, in tribal cultures, initiates a metamorphosis wherein the person's presence does not "vanish" from the sensible world (where would it go?) but rather remains as an animating force within the vastness of the landscape, whether subtly, in the wind, or more visibly, in animal form, or even as the eruptive, ever to be appeased, wrath of the volcano. "Ancestor worship," in its myriad forms, then, is ultimately another mode of attentiveness to nonhuman nature; it signifies not so much an awe or reverence of human powers, but rather a reverence for those forms that awareness takes when it is *not* in human form, when the familiar human embodiment dies and decays to become part of the encompassing cosmos.

This cycling of the human back into the larger world ensures that the other forms of experience that we encounter—whether ants, or willow trees, or clouds—are never absolutely alien to ourselves. Despite the obvious differences in shape, and ability, and style of being, they remain at least distantly familiar, even familial. It is, paradoxically, this perceived kinship or consanguinity that renders the difference, or otherness, so eerily potent.

Several months after my arrival in Bali, I left the village in which I was staying to visit one of the pre-Hindu sites on the island. I arrived on my bicycle early in the afternoon, after the bus carrying tourists from the coast had departed. A flight of steps took me down into a lush, emerald valley, lined by cliffs on either side, awash with the speech of the river and the sighing of the wind through high, unharvested grasses. On a small bridge crossing the river I met an old woman carrying a wide basket on her head and holding the hand of a little, shy child; the woman grinned at me with the red, toothless smile of a betel-nut chewer. On the far side of the river I stood in front of a great moss-covered complex of passageways, rooms, and courtyards carved by hand out of the black volcanic rock.

I noticed, at a bend in the canyon downstream, a further series of caves carved into the cliffs. These appeared more isolated and remote, unattended by any footpath I could discern. I set out through the grasses to explore them. This proved much more difficult than I anticipated, but after getting lost in the tall grasses, and fording the river three times, I at last found myself beneath the caves. A short scramble up the rock wall brought me to the mouth of one of them, and I entered on my hands and knees. It was a wide but low opening, perhaps only four feet high, and the interior receded only about five or six feet into the cliff. The floor and walls were covered with mosses, painting the cave with green patterns and softening the harshness of the rock; the place, despite its small size—or perhaps because of it—had an air of great friendliness. I climbed to two other caves, each about the same size, but then felt drawn back to the first one, to sit cross-legged on the cushioning moss and gaze out across the emerald canyon. It was quiet inside, a kind of intimate sanctuary hewn into the stone. I began to explore the rich resonance of the enclosure, first just humming, then intoning a simple chant taught to me by a balian some days before. I was delighted by the overtones that the cave added to my voice, and sat there singing for a long while. I did not notice the change in the wind outside, or the cloud shadows darkening the valley, until the rains broke—suddenly and with great force. The first storm of the monsoon!

I had experienced only slight rains on the island before then, and was startled by the torrential downpour now sending stones tumbling along the cliffs, building puddles and then ponds in the green landscape below, swelling the river. There was no question of returning home—I would be unable to make my way back through the flood to the valley's entrance. And so, thankful for the shelter, I recrossed my legs to wait out the storm. Before long the rivulets falling along the cliff above gathered themselves into streams, and two small waterfalls cascaded across the cave's mouth. Soon I was looking into a solid curtain of water, thin in some places, where the canyon's image flickered unsteadily, and thickly rushing in others. My senses were all but overcome by the wild beauty of the cascade and by the roar of sound, my body trembling inwardly at the weird sense of being sealed into my hiding place.

And then, in the midst of all this tumult, I noticed a small, delicate activity. Just in front of me, and only an inch or two to my side of the torrent, a spider was climbing a thin thread stretched across the mouth of the cave. As I watched, it anchored another thread to the top of the opening, then slipped back along the first thread and joined the two at a point about midway between the roof and the floor. I lost sight of the spider then, and for a while it seemed that it had vanished, thread and all, until my focus rediscovered it. Two more threads now radiated from the center to the floor, and then another; soon the spider began to swing between these as on a circular trellis, trailing an ever-lengthening thread which it affixed to each radiating run as it moved from one to the next, spiraling outward. The spider seemed wholly undaunted by the tumult of waters spilling past it, although every now and then it broke off its spiral dance and climbed to the roof or the floor to tug on the radii there, assuring the tautness of the threads, then crawled back to where it left off. Whenever I lost the correct focus, I waited to catch sight of the spinning arachnid, and then let its dancing form gradually draw the lineaments of the web back into visibility, tying my focus into each new knot of silk as it moved, weaving my gaze into the ever-deepening pattern.

And then, abruptly, my vision snagged on a strange incongruity: another thread slanted across the web, neither radiating nor spiraling from the central juncture, violating the symmetry. As I followed it with my eyes, pondering its purpose in the overall pattern, I began to realize that it was on a different plane from the rest of the web, for the web slipped out of focus whenever this new line became clearer. I soon saw that it led to its own center, about twelve inches to the right of the first, another nexus of forces from which several threads stretched to the floor and the ceiling. And then I saw that there was a *different* spider spinning this web, testing its tautness by dancing around it like the first, now setting the silken cross weaves around the nodal point and winding outward. The two spiders spun independently of each other, but to my eyes they wove a single intersecting pattern. This widening of my gaze soon disclosed yet another spider spiraling in the cave's mouth, and suddenly I realized that there were *many* overlapping webs coming into being, radiating out at different rhythms from myriad centers poised—some higher, some lower, some minutely closer to my eyes and some farther—between the stone above and the stone below.

I sat stunned and mesmerized before this ever-complexifying expanse of living patterns upon patterns, my gaze drawn like a breath into one converging group of lines, then breathed out into open space, then drawn down into another convergence. The curtain of water had become utterly silent—I tried at one point to hear it, but could not. My senses were entranced. I had the distinct impression that I was watching the universe being born, galaxy upon galaxy.

Night filled the cave with darkness. The rain had not stopped. Yet, strangely, I felt neither cold nor hungry—only remarkably peaceful and at home. Stretching out upon the moist, mossy floor near the back of the cave, I slept.

When I awoke, the sun was staring into the canyon, the grasses below rippling with bright blues and greens. I could see no trace of the webs, nor their weavers. Thinking that they were invisible to my eyes without the curtain of water behind them, I felt carefully with my hands around and through the mouth of the cave. But the webs were gone. I climbed down to the river and washed, then hiked across and out of the canyon to where my cycle was drying in the sun, and headed back to my own valley.

I have never, since that time, been able to encounter a spider without feeling a great strangeness and awe. To be sure, insects and spiders are not the only powers, or even central presences, in the Indonesian universe. But they were *my* introduction to the spirits, to the magic afoot in the land. It was from them that I first learned of the intelligence that lurks in nonhuman nature, the ability that an alien form of sentience has to echo one's own, to instill a reverberation in oneself that temporarily shatters habitual ways of seeing and feeling, leaving one open to a world all alive, awake, and aware. It was from such small beings that my senses first learned of the countless worlds within worlds that spin in the depths of this world that we commonly inhabit, and from them that I learned that my body could, with practice, enter sensorially into these dimensions. The precise and minuscule craft of the spiders had so honed and focused my awareness that the very web-work of the universe, of which my own flesh was a part, seemed to be being spun by their arcane art. I have already spoken of the ants, and of the fireflies, whose sensory likeness to the lights in the night sky had taught me the fickleness of gravity. The long and cyclical trance that we call malaria was also brought to me by insects, in this case mosquitoes, and I lived for three weeks in a feverish state of shivers, sweat, and visions.

I had rarely before paid much attention to the natural world. But my exposure to traditional magicians and seers was shifting my senses; I became increasingly susceptible to the solicitations of nonhuman things. In the course of struggling to decipher the magicians' odd gestures or to fathom their constant spoken references to powers unseen and unheard, I began to *see* and to *hear* in a manner I never had before. When a magician spoke of a power or "presence" lingering in the corner of his house, I learned to notice the ray of sunlight that was then pouring through a chink in the roof, illuminating a column of drifting dust, and to realize that that column of light was indeed a power, influencing the air currents by its warmth, and indeed influencing the whole mood of the room; although I had not consciously seen it before, it had already been structuring my experience. My ears began to attend, in a new way, to the songs of birds—no longer just a melodic background to human speech, but meaningful speech in its own right, responding to and commenting on events in the surrounding earth. I became a student of subtle differences: the way a breeze may flutter a single leaf on a whole tree, leaving the other leaves silent and unmoved (had not that leaf, then, been brushed by a magic?); or the way the intensity of the sun's heat expresses itself in the precise rhythm of the crickets. Walking along the dirt paths, I learned to slow my pace in order to *feel* the difference between one nearby hill and the next, or to taste the presence of a particular field at a certain time of day when, as I had been told by a local *dukun*, the place had a special power and proffered unique gifts. It was a power communicated to my senses by the way the shadows of the trees fell at that hour, and by smells that only then lingered in the tops of the grasses without being wafted away by the wind, and other elements I could only isolate after many days of stopping and listening.

And gradually, then, other animals began to intercept me in my wanderings, as if some quality in my posture or the rhythm of my breathing had disarmed their wariness; I would find myself face-to-face with monkeys, and with large lizards that did not slither away when I spoke, but leaned forward in apparent curiosity. In rural Java, I often noticed monkeys accompanying me in the branches overhead, and ravens walked toward me on the road, croaking. While at Pangandaran, a nature preserve on a peninsula jutting out from the south coast of Java ("a place of many spirits," I was told by nearby fishermen), I stepped out from a clutch of trees and found myself looking into the face of one of the rare and beautiful bison that exist only on that island. Our eyes locked. When it snorted, I snorted back; when it shifted its shoulders, I shifted my stance; when I tossed my head, it tossed *its* head in reply. I found myself caught in a nonverbal conversation with this Other, a gestural duet with which my conscious awareness had very little to do. It was as if my body in its actions was suddenly being motivated by a wisdom older than my thinking mind, as though it was held and moved by a logos, deeper than words, spoken by the Other's body, the trees, and the stony ground on which we stood.

Although the Indonesian islands are home to an astonishing diversity of birds, it was only when I went to study among the Sherpa people of the high Himalayas that I was truly initiated into the avian world. The Himalayas are young mountains, their peaks not yet rounded by the endless action of wind and ice, and so the primary dimension of the visible landscape is overwhelmingly vertical. Even in the high ridges one seldom attains a view of a distant horizon; instead one's vision is deflected upward by the steep face of the next mountain. The whole land has surged skyward in a manner still evident in the lines and furrows of the mountain walls, and this ancient dynamism readily communicates itself to the sensing body.

In such a world those who dwell and soar in the sky are the primary powers. They alone move easily in such a zone, swooping downward to become a speck near the valley floor, or spiraling into the heights on invisible currents. The wingeds, alone, carry the immediate knowledge of what is unfolding on the far side of the next ridge, and hence it is only by watching them that one can be kept apprised of climatic changes in the offing, as well as of subtle shifts in the flow and density of air currents in one's own valley. Several of the shamans that I met in Nepal had birds as their close familiars. Ravens are constant commentators on village affairs. The smaller, flocking birds perform aerobatics in unison over the village rooftops, twisting and swerving in a perfect sympathy of motion, the whole flock appearing like a magic banner that floats and flaps on air currents over the village, then descends in a heap, only to be carried aloft by the wind a moment later, rippling and swelling.

For some time I visited a Sherpa *dzankri* whose rock home was built into one of the steep mountainsides of the Khumbu region in Nepal. On one of our walks along the narrow cliff trails that wind around the mountain, the *dzankri* pointed out to me a certain boulder, jutting out from the cliff, on which he had "danced" before attempting some especially difficult cures. I recognized the boulder several days later when hiking back down toward the *dzankri's* home from the upper yak pastures, and I climbed onto the rock, not to dance but to ponder the pale white and red lichens that gave life to its surface, and to rest. Across the dry valley, two lammergeier condors floated between gleaming, snow-covered peaks. It was a ringing blue Himalayan day, clear as a bell. After a few moments I took a silver coin out of my pocket and aimlessly began a simple sleight-of-hand exercise, rolling the coin over the knuckles of my right hand. I had taken to practicing this somewhat monotonous exercise in response to the endless flicking of prayer-beads by the older Sherpas, a practice usually accompanied by a repetitively chanted prayer: "*Om Mani Padme Hum*" (O the Jewel in the Lotus). But there was no prayer accompanying my revolving coin, aside from my quiet breathing and the dazzling sunlight. I noticed that one of the two condors in the distance had swerved away from its partner and was now floating over the valley, wings outstretched. As I watched it grow larger, I realized, with some delight, that it was heading in my general direction; I stopped rolling the coin and stared. Yet just then the lammergeier halted in its flight, motionless for a moment against the peaks, then swerved around and headed back toward its partner in the distance. Disappointed, I took up the coin and began rolling it along my knuckles once again, its silver surface catching the sunlight as it turned, reflecting the rays back into the sky. Instantly, the condor swung out from its path and began soaring back in a wide arc. Once again, I watched its shape grow larger. As the great size of the bird became apparent, I felt my skin begin to crawl and come alive, like a swarm of bees all in motion, and a humming grew loud in my ears. The coin continued rolling along my fingers. The creature loomed larger, and larger still, until, suddenly, it was there—an immense silhouette hovering just above my head, huge wing feathers rustling ever so slightly as they mastered the breeze. My fingers were frozen, unable to move; the coin dropped out of my hand. And then I felt myself stripped naked by an alien gaze infinitely more lucid and precise than my own. I do not know for how long I was transfixed, only that I felt the air streaming past naked knees and heard the wind whispering in my feathers long after the Visitor had departed.

I returned to North America . . . excited by the new sensibilities that had stirred in me—my newfound awareness of a more-than-human world, of the great potency of the land, and particularly of the keen intelligence of other animals, large and small, whose lives and cultures interpenetrate our own. I startled neighbors by chattering with squirrels, who swiftly climbed down the trunks of their trees and across lawns to banter with me, or by gazing for hours on end at a heron fishing in a nearby estuary, or at gulls opening clams by dropping them from a height onto the rocks along the beach.

Yet, very gradually, I began to lose my sense of the animals' own awareness. The gulls' technique for breaking open the clams began to appear as a largely automatic behavior, and I could not easily feel the attention that they must bring to each new shell. Perhaps each shell was entirely the same as the last, and *no* spontaneous attention was really necessary. . . .

I found myself now observing the heron from outside its world, noting with interest its careful high-stepping walk and the sudden dart of its beak into the water, but no longer feeling its tensed yet poised alertness with my own muscles. And, strangely, the suburban squirrels no longer responded to my chittering calls. Although I wished to, I could no longer focus my awareness on engaging in their world as I had so easily done a few weeks earlier, for my attention was quickly deflected by internal, verbal deliberations of one sort or another—by a conversation I now seemed to carry on entirely within myself. The squirrels had no part in this conversation.

It became increasingly apparent, from books and articles and discussions with various people, that other animals were not as awake and aware as I had assumed, that they lacked any real language and hence the possibility of thought, and that even their seemingly spontaneous responses to the world around them were largely "programmed" behaviors, "coded" in the genetic material now being mapped by biologists. Indeed, the more I spoke *about* other animals, the less possible it became to speak *to* them. I gradually came to discern that there was no common ground between the unlimited human intellect and the limited sentience of other animals, no medium through which we and they might communicate with and reciprocate one another.

As the expressive and sentient landscape slowly faded behind my more exclusively human concerns, threatening to become little more than an illusion or fantasy, I began to feel—particularly in my chest and abdomen—as though I were being cut off from vital sources of nourishment. I was indeed reacclimating to my own culture, becoming more attuned to its styles of discourse and interaction, yet my bodily senses seemed to be losing their acuteness, becoming less awake to subtle changes and patterns. The thrumming of crickets, and even the songs of the local blackbirds, readily faded from my awareness after a few moments, and it was only by an effort of will that I could bring them back into the perceptual field. The flight of sparrows and of dragonflies no longer sustained my focus very long, if indeed they gained my attention at all. My skin quit registering the various changes in the breeze, and smells seemed to have faded from the world almost entirely, my nose waking up only once or twice a day, perhaps while cooking, or when taking out the garbage.

In Nepal, the air had been filled with smells—whether in the towns, where burning incense combined with the aromas of roasting meats and honeyed pastries and fruits for trade in the open market, and the stench of organic refuse rotting in the ravines, and sometimes of corpses being cremated by the river; or in the high mountains, where the wind carried the whiffs of countless wildflowers, and of the newly turned earth outside the villages where the fragrant dung of yaks was drying in round patties on the outer walls of the houses, to be used, when dry, as fuel for the household fires, and where smoke from those many home fires always mingled in the outside air. And sounds as well: the chants of aspiring monks and adepts blended with the ringing of prayer bells on near and distant slopes, accompanied by the raucous croaks of ravens, and the sigh of the wind pouring over the passes, and the flapping of prayer flags, and the distant hush of the river cascading through the far-below gorge.

There the air was a thick and richly textured presence, filled with invisible but nonetheless tactile, olfactory, and audible influences. In the United States, however, the air seemed thin and void of substance or influence. It was not, here, a sensuous medium—the felt matrix of our breath and the breath of the other animals and plants and soils—but was merely an absence, and indeed was constantly referred to in everyday discourse as mere empty space. Hence, in America I found myself lingering near wood fires and even garbage dumps—much to the dismay of my friends—for only such an intensity of smells served to remind my body of its im-

mersion in an enveloping medium, and with this experience of being immersed in a world of influences came a host of body memories from my year among the shamans and village people of rural
Asia.

Anthropology's inability to discern the Shaman's allegiance to
nonhuman nature has led to a curious circumstance in the "developed world" today, where many persons in search of spiritual understanding are enrolling in workshops concerned with "shamanic"
methods of personal discovery and revelation. Psychotherapists and
some physicians have begun to specialize in "shamanic healing
techniques." "Shamanism" has thus come to connote an alternative
form of therapy; the emphasis, among these new practitioners of
popular shamanism, is on personal insight and curing. These are
noble aims, to be sure, yet they are secondary to, and derivative
from, the primary role of the indigenous shaman, a role that cannot be fulfilled without long and sustained exposure to wild nature, to its patterns and vicissitudes. Mimicking the indigenous
shaman's curative methods without his intimate knowledge of the
wider natural community cannot, if I am correct, do anything
more than trade certain symptoms for others, or shift the locus of
dis-ease from place to place within the human community. For the
source of stress lies in the relation *between* the human community
and the natural landscape.

Western industrial society, of course, with its massive scale and
hugely centralized economy, can hardly be seen in relation to any
particular landscape or ecosystem; the more-than-human ecology
with which it is directly engaged is the biosphere itself. Sadly, our
culture's relation to the earthly biosphere can in no way be considered a reciprocal or balanced one: with thousands of acres of
nonregenerating forest disappearing every hour, and hundreds of
our fellow species becoming extinct each month as a result of our
civilization's excesses, we can hardly be surprised by the amount
of epidemic illness in our culture, from increasingly severe im-
mune dysfunctions and cancers, to widespread psychological dis-
tress, depression, and ever more frequent suicides, to the acceler-

ating number of household killings and mass murders committed
for no apparent reason by otherwise coherent individuals.

From an animistic perspective, the clearest source of all this dis-
tress, both physical and psychological, lies in the aforementioned
violence needlessly perpetrated by our civilization on the ecology
of the planet; only by alleviating the latter will we be able to heal
the former. While this may sound at first like a simple statement
of faith, it makes eminent and obvious sense as soon as we acknowledge our thorough dependence upon the countless other organisms with whom we have evolved. Caught up in a mass of abstractions, our attention hypnotized by a host of human-made
technologies that only reflect us back to ourselves, it is all too easy
for us to forget our carnal inherence in a more-than-human matrix
of sensations and sensibilities.

Our bodies have formed themselves in delicate reciprocity with
the manifold textures, sounds, and shapes of an animate earth—
our eyes have evolved in subtle interaction with *other* eyes, as our
ears are attuned by their very structure to the howling of wolves
and the honking of geese. To shut ourselves off from these other
voices, to continue by our lifestyles to condemn these other sensibilities to the oblivion of extinction, is to rob our own senses of
their integrity, and to rob our minds of their coherence. We are
human only in contact, and conviviality, with what is not human.
Only in reciprocity with what is Other do we begin to heal ourselves.

2

Is It Too Late?

Anthony Weston

Desolation

The amazement of the first Europeans at America now amazes *us*; it breaks our hearts. The first explorers smelled the land before they even saw it. In 1524, Verrazano reported smelling cedars a hundred leagues from land. Others told of sailing through vast beds of floating flowers. Ducks, turkey, deer, lynx greeted them in unimaginable abundance. Whales so crowded the waters that they were navigational hazards. Cape Cod was named for its cod-clotted waters. Salmon ran thick in every Atlantic river from Labrador to the Hudson. Lobsters were so common that they were used throughout New England for potato fertilizer, pig food, fish bait, and to feed the British navy five or six times a week until seamen's riots limited the weekly lobster dinners to four.

All gone, now. Islands once packed shore to shore with walrus or seals or nesting seabirds are now empty. The spearbill, just for one of untold possible examples, is extinct, slaughtered for its eggs or for cod bait, thousands burnt alive for fuel to melt down others' fat. Passenger pigeons who once flew in hundred-mile long rivers over the prairies, so thick they blotted out the sun, perhaps the most abundant bird species ever to live on Earth, no longer exist at all. Buffalo numbered perhaps seventy million in North America prior to Europeans' arrival on the prairies; they were reduced to only a few hundred less than a century later, and even now are regularly shot if they dare to wander out of Yellowstone or their few other sanctuaries. Some subspecies, now extinct, were larger than elephants. Pilot whales, giving birth and giving suck in shallow coastal bays, were shot with cannons or driven onto the beaches and left to die, then cut up, often still alive, sometimes after two or three days of slowly collapsing under their own weight waiting for the butchers to finish their companions. In the heyday of deep-sea whaling it was no longer live whales who were navigational hazards, but *dead* ones, stripped of their blubber and cast adrift.[1]

The destruction goes on. Vast tracts of rainforest, harboring untold species that have not even been counted, are in flames at this moment. Spotted owl, willow flycatcher, scarlet tanager, lynx, bobcat, rhinoceros, elephant: all are or will be endangered. Three dozen species have gone extinct in the United States in the last decade just waiting for Endangered Species Act designation. Fully one quarter of the twenty thousand native plants in America are threatened with extinction. Blue whales, the largest living creatures on earth, have been reduced from a half a million in Antarctic waters alone to between 600 and 3,000 worldwide.[2] The use of whales and other cetaceans for target-practice (machine-gunning, depth charging, ramming) by the world's navies continues to be so common that NATO commanders are surprised when there are protests. Old-growth forest in Oregon and Washington, much of it public land, may be entirely clear-cut in a generation. Almost as much land in the lower forty-eight states is dedicated to roads, rights of way, and parking lots as is designated wilderness. An area the size of Indiana is occupied just by lawns. Every major city in the country now publishes daily air pollution indices (is it safe to breathe the air? as if we had a choice), with local spice added too: the drinking water in Des Moines, Iowa, for instance, is so contaminated by leachate from fertilizers that nitrate levels are featured in the news as well. The ground water on the Wisconsin prairie where I grew up now contains so much nitrate that it could kill babies. My own children, visiting their grandfather in the country, have to drink bottled water.

Only a little less familiar are slower and more indirect processes of destruction. Many species, even those apparently well-protected, are dying out slowly as their habitat is destroyed. Migratory songbirds like tanagers are threatened by loss of habitat at both ends (they winter in the rainforests) as well as by pesticides at this end. Nitrates too, no doubt (for what do *their* babies drink?). The young produced by the twenty-seven previously wild California condors, captured and transferred to breeding programs in zoos in 1987, may or may not reacclimate to the wild, and all of the causes of their original decline still remain: loss of habitat, poison (lead shot in dead animals not recovered by hunters), high-energy powerlines. Of an unknown number of subspecies of wolves in North America (Edward Goldman distinguished twenty-three in 1945), seven are extinct (like the pure white Newfoundland wolf, *Canis lupus beothucus*, named after Newfoundland's original human inhabitants, the Beothuk Indians—likewise extinct) and the rest, surviving after a fashion even under the pressure of unparalleled destruction and dislocation, seem to have interbred with each other and with dogs and coyotes to the point that only two or three subspecies remain. Bats that once flew in huge clouds from caves have been killed by the millions by one person sealing an entrance or destroying their hearing with a bomb; yet bats are the pollinators for a wide variety of crucial species. Blue whales move in groups that may be too isolated to interbreed. Even with a global population in the thousands, the species may be dying out. North Atlantic Right Whales, after not being hunted for fifty years, still number only a few hundred.[3]

Inconceivable pain, horror and holocaust lie behind these briefly listed facts. "Holocaust" literally means "cast whole into the fire"— think of the burning rainforests. Loss to untold numbers of animals, to the future of the whole earth, and to those human peoples who have lived close to those animals and to the land. Aboriginal peo-

ples worldwide are being uprooted or killed off, by outright murder or starvation, dislocation, disease. Often the vanishing is quiet,
hidden and of no concern to the dominant civilization that moves
in to fill the vacuum, or just lets the ghost towns (ghost cultures,
ghost languages, ghost possibilities) crumble into dust. Inuit communities, visited once by a whaling ship, died to a person the next
winter. Indian genocide through smallpox-infected blankets was
deliberate U.S. government policy. Similar things go on today in
the Amazon. The extermination of the buffalo was likewise a deliberate attack on the Plains Indians. Often the natives are the only
humans who can give a fully situated and *live* voice to the loss of
the animals. A Nisqually Indian spokesperson, speaking of salmon—
the being who, for all the Northwest Indians, "gathers" the landscape, but now almost completely exterminated, blocked from their
breeding grounds by hydroelectric dams or chopped to bits in their
turbines—tells us that the Nisqually see salmon bleeding out of
light bulbs. We are the blind ones, seeing only the light.[4]

And if this were not enough, the threats on the horizon dwarf
all that has gone before. Most of these are so familiar that we can
invoke them merely by the briefest reference. Global warming,
brought on chiefly by the burning of fossil fuels in cars and power
plants, and its threat of coastal inundation, superhurricanes, and
drastic and unpredictable climate changes. Ozone depletion, starting with the infamous hole over the Antarctic and potentially continuing worldwide. Already more of the sun's ultraviolet rays reach
Earth's surface than a decade ago: Australia, on the edge of the
Antarctic hole, has seen skin cancers treble. Everywhere too lie
hidden dangers, "risk multipliers," processes that intensify other
processes: global warming may speed up the decomposition of the
dead organic matter, for example, that now lies on forest floors,
flooding the atmosphere with vast new quantities of carbon dioxide and accelerating further warming. Then there are other and
more technological nightmares: nuclear war or meltdown, massive
chemical spills, the Chernobyls and Bhopals still in our future,
perhaps down the street; some sort of genetically engineered organism run amok; assorted other cataclysms. No end of nightmares. I cannot bear to begin to tell my children.

No wonder we feel ourselves in a kind of endgame: that it is already "too late." Every year I find my students more and more knowledgeable about environmental issues, and every year more fatalistic. The trends seem to be connected. There is nothing to be done, they say. The jig is up.

And Yet . . .

And yet . . . the very fact that we have awakened to all of this is also, perversely enough, a sign of progress. Only a first sign, perhaps, but a sign nonetheless. The list of cataclysms, past and possible, is no longer a surprise—*and this is itself a hopeful sign.* No problem will be solved without the awareness that it *is* a problem. For the first time in centuries, we are now aware, truly and inescapably aware, that we ourselves are *at stake with* the larger living world. And the speed with which this awareness has grown on us, seen from any historical distance, is astonishing too. Thirty years ago the United States did not even have an Environmental Protection Agency. Today it is a Cabinet-level department, its enabling legislation together with the Clean Air and Clean Water Acts and the Endangered Species Act impose direct economic costs on the scale of $125 billion a year, and three-quarters of the public approves.[5]

More profound changes lie ahead—changes that would have seemed impossible, unimaginable, a decade or two ago. Tighter pollution control, habitat protection, limits to the factory farming of animals, usable mass transit, sustainable agriculture, an enhanced Endangered Species Act, and on and on: all of this is on the agenda. No doubt there will be immense struggles over what shape they will take and how fast they will come, but that they are coming is not in serious question. This too we know; it is no surprise. Step back from the immediate struggles and a larger movement becomes more visible. Systematic recycling is growing on everyone: municipalities can no longer afford to dispose of the current volume of garbage. There are no more landfills. The next step is to rethink the manufacture of so many "disposable" items in the first place. It is far cheaper and more sensible to plan products so that recycling is easy—planners call this *"precycling"*—than to pay attention only after the things are made, after they have been designed and manufactured with no attention to anything but the manufacturer's ease and profits. Even better, eliminate disposables entirely. Make containers to last through many reuses. Or make them edible. Or fertilize the garden with them, as we already do with some newsprint. We speak here of the growth industries of the first part of the twenty-first century.

Even the power companies are getting the message. It is far cheaper to conserve electricity—creating what Amory Lovins calls *"negawatts"*—than to build new generating capacity, especially when the environmental costs of electricity generation are factored in, but even if they are not. Of necessity, it is happening. No new nuclear reactors, and few large power plants of any sort, are currently under construction in America. Hardly any dams. Willynilly, ecological thinking has arrived.

Other changes proceed apace. Some of my friends live in a cooperatively owned and managed development, the houses clumped around the periphery with shared open, natural, and quiet space at the center. New developments all over the country are being similarly designed. Urban creeks, buried underground or channeled straight and narrow, are coming back out of their confinement. There is now an entire organization called the National Coalition to Restore Urban Waters. Restored wetlands are replacing shopping centers. Vegetarianism, on the utter fringe twenty years ago, is now familiar enough that people feel they need excuses for eating meat. Organic foods are common enough that conventional food producers, poor things, are starting to worry that people may actually "overreact" to the latest pesticide scare and switch. Nearly half of all U.S. supermarkets now carry some organic produce. And organic farming is explicitly conceived as a commitment to the health of soils and ecosystems, not just to the food consumers at the other end. It represents a new *vision* of farming as well as of food.

So change is the order of the day, and not all the change is downhill. Ten minutes from my urban home there is thick forest where eighty years ago farmland was eroded and abandoned. Succession into hardwoods is starting. Who would have dreamed it a century ago? Now we have before us proposals to turn vast, sparsely settled and overfarmed parts of the Great Plains into "Buffalo Commons": back into open prairie, restoring the native plants and animals, and making space again for the native peoples too. Already throughout New England the wilderness trails cross old stone walls, the walls of farms two hundred years ago. *Wilderness* now. The face of the land may be utterly transformed again in another century.

This is what I do tell my children. I want them to awaken to a world understood ecologically: to a vision of things according to which, for example, buying organic food is about the health of the land and of the people who work it; hauling the recycling to the curb (or, better, doing our own freelance "pre-cycling") is about stewardship and creativity; and walking the woods is partly about regeneration. Time enough later for the horrors: first the new vision, the new century in a new key. A new *millennium,* maybe, in a new key. Perhaps, slowly, the tide is turning.

Our Fatalism

It may come as a surprise that fatalism is nothing new. Actually, people have been lamenting that "it is too late" for millennia, almost from the very morning of the world. The great Greek philosophers were obsessed with the theme of decay and decline. Christianity reflects a pessimism so deep that the only sources of hope have to be moved outside of the world entirely. So now ecology is supposed to offer us a post-Christian, scientific kind of fatalism: the Earth itself is in decline, life itself is imperilled. A new kind of necessity, we are told, a new wheel of fate upon which we are once again pinned, has now been discovered.

There is of course evidence for this view, just as there is evidence for many kinds of pessimism. Yet there is a strange and striking way in which this fatalism treats the Earth as though its possibilities are exhausted—*at the very moment when we are finally*

*supposed to have learned that the Earth has infinitely more possibilities
than we ever imagined!* Is it not a little odd to give up on the Earth
just as we discover how fantastically intricate and varied and mysterious the Earth (even a single salt marsh, even a single *duck*) really is? The truth is that *we barely know this place.* How can we
know enough to give up on it?

In the 1960s we were told, on the best of evidence, that a billion or more people would starve to death by 1990, because population had irrevocably outgrown food supplies. Since then, population growth has slowed down while food supplies have increased,
and although widespread starvation and malnutrition exist, they
are arguably the result of political causes, not (primarily) population explosion. The "Population Bomb" didn't go off. Yet we were
already told, in the 1960s, that it was "too late." It wasn't. Do we
know it is "too late" now? Are we sure? Prediction is dangerous,
as E. F. Schumacher once quipped, especially about the future.

Global warming is the disaster forecast of choice in the 1990s.
But is prediction any more secure here? Almost all of our theories
about the Earth are guesswork, far more uncertain than we usually
admit, especially as the issues become more global, more dependent on projections, computer models, and ecological theories.
Even the basic data are woefully incomplete. Surface temperatures
around the world do appear to be increasing. Satellite data for the
lower atmosphere are more precise but show a less clear pattern,
but then again measurements from space have only been taken for
seventeen years, not long enough to reliably judge trends. It is also
not clear that the reported increases don't have other explanations.
What determines Earth's temperature is almost incomprehensibly
complicated. Variations in solar activity and even small variations
in the tilt of Earth's orbit are crucial. Meanwhile none of our theories explain some of the actual empirical phenomena, such as the
marked lack of warming at the poles. It is not even clear that
global warming, if it really does happen, will raise the oceans. The
Antarctic glacial shield was formed thirty million years ago and
has withstood much warmer temperatures that are predicted at
present. Since a warmer globe will experience more evaporation
and precipitation, there will be more snow on the polar ice caps,

thickening them: some recent estimates suggest that the sea may actually *drop*.[6]

I do *not* mean (this had better be said emphatically: I do NOT mean) that we may as well go on as we are going. When we don't understand the dynamics of a process we depend upon, the only sane policy is not to interfere with it. Heedlessly dumping billions of tons of pollutants into the air and water is like an ignorant patient on life support polluting his own intravenous solution, flipping dials on the respirator, pulling out tubes and jacks and plugging them into other outlets. You don't have to understand the process to know that messing with the system is, to say the least, unwise. What goes around quite probably will come around. In fact it is precisely *because* we don't understand the process that it *is* so unwise. Rather than requiring proof that our present course is disastrous before we change it, then, we ought to require proof that it is *not* disastrous before we embark on it. Anything else is sheer stupidity—probably based in turn upon sheer *cu*pidity.

Yet a fundamental uncertainty remains, and may always remain. So what makes caution and forbearance necessary is also precisely what also opens the space, once again, for hope. If the Earth is truly mysterious, if so many of her intricacies flow onward above or below or beyond us, then, even though she is obviously pained, there is no way that we can say that it is now "too late." Or *ever* "too late."

Believing in the latest disaster scenario, the latest fatalism, ought not to be the litmus test of environmentalism. We ought to honor this Earth in her *elusiveness*. This Earth eludes the pesticide makers, who find each new chemical shortly greeted by resistant insects that are even more destructive and harder to control. It eludes the nuclear industry's planners, unable to find any place, even deep within the earth, to entomb nuclear wastes, because even the rocks move. It eludes the paleontologists, who estimate that, even now, we have discovered at best 1 percent—*1 percent!*—of dinosaur species. Who knows what awaits us, buried in the rocks? It eludes the computer modellers, who still, apparently, even now, have no idea where a billion tons of carbon dioxide—a seventh or more of the total dumped into the atmosphere from human sources—goes every year, though some of them confidently go on to predict, or deny, global warming anyway. Wholly unbelievable lifeforms may yet find their way out of the rainforests, as some already have (Who knows: an AIDS virus in reverse?). Or out of the oceans, like the coelacanth, a prehistoric, dragon-like fish, believed extinct for seventy million years, rediscovered off South Africa in 1938, brought to attention by an eccentric British scientist—a story that sounds like something out of a Gary Larson cartoon, except that it is true.[7]

Dinosaurs are not extinct either, by the way: it turns out they evolved into birds. The littlest and flightiest from the biggest and the most plodding. Whales evolved from a dog-like land mammal who came from the sea and then, a few hundred million years later, went back. And nothing could be stranger than what is already around us: fungi (closer cousins than we thought, research suggests[8]), flamingos, insects of all sorts (watch the movie *Microcosmos* for an amazing eyeful: slugs making love, delicately and languidly; the nativity of a mosquito . . .), elephants rumbling to each other across the savannahs (it turns out they communicate by infrasound, like the rumble of cathedral organ pipes, almost too low for us to hear). What new coelacanths must still swim below the waters we think we know so well! The proverbial grass cracks all pavements. The jig is up? Not likely.

Smash the Tube

I think the current fatalism comes primarily from watching the news. At least this much is true: most of us know what we know about nature from the media. On the one hand, I suppose, this is wonderful: perhaps only with such media can consciousness shift dramatically toward an ecological vision. But there is another and darker side too. In "the news" especially, a very specific concept of nature is at play.

Nature on TV is not something we directly experience. First and obviously, of course, it is *on* television: we are experiencing it only in a "mediated" way. In a darkened room, a tube flickers; we sit, passive, reduced to our eyes. But there is another reason too. In the time-honored fashion of television "coverage," nature is treated as such a large and complex system that we can only learn about it from the experts, the talking heads, just like the national debt or genetic engineering or politics in Russia. Or else nature makes news when trouble comes, so we have also learned to picture it as trouble-prone, threatening. For us North Carolinians, it seems that "nature" on the tube is primarily the place that hurricanes come from.

Thus television's nature is pictured as, in general, *somewhere else*, which is *why* we need experts to tell us about it, and why it can leave us so uneasy, vaguely but only vaguely sensing storms gathering over the horizon. And this in turn is part of why it seems so easy and tempting to abandon hope: because we feel no connection to television's Earth in our bones, and also, therefore, are offered no sense that it is something we can directly affect. The implicit message is: there is nothing to do, it is indeed too late, all we can do is sit back and watch the unfolding disaster, between the ads for toothpastes and senators, on TV.

But the real news is that *we are all part of nature*. We humans do not stand apart, and nature does not stand apart from us. We all stand in dynamic interaction. The system is open-ended and far more resourceful, as I have been arguing, than we usually suspect: and it is also *right here*, right next to us, and we too are part of it. Nature is us.

From this perspective, all of the familiar aspects of "crisis" are still on the scene, but the overall picture is nonetheless strikingly different. Of course there are deep threats to certain parts of nature, especially the wilder and more intricate systems. Of course we may be disequilibrating some delicate global feedback mechanisms that could eventually disrupt the entire biosphere. Of course we must take these threats seriously—and again, emphatically, nothing I say here is meant to diminish them. But we also must resist the reduction or confinement of "nature" to *just that*—as if everything else were confined to some other, lesser, incomplete sort of reality, *off the screen*, as it were. Another TV image, this time from the advertisements: "nature" is something pristine far away, like a vacation destination. Never mind that if many of us vacation there it will no longer be pristine. Never mind that this entire conception is a product of the airline and hotel industries' need to generate demand for their services. All the same, it has now become reality for all too many of us.

This is a spell we need to break—and all it takes, in truth, is looking up. Look at the world, *really look*; listen and smell and touch. We do not live apart. A cat hunches over this very desk as I write, bats hang under your eaves, 100 million monarch butterflies migrate, every year, up to 4,000 miles, from all over North America, to winter on the California coast, following what nature writer John Hay calls "nature's great headings," invisible to us. All of this is going on around us, right now, all the time. The same dainty-looking butterflies (a lovely children's tale calls them "flutter-bys") welcome the spring in all our gardens. There are wild worlds under our feet: the billion or so microbes in a handful of garden soil, and beneath them bacteria living in rocks up to three miles below Earth's surface: the latest theory, in fact, is that life on Earth may have originated in the interior, and that subterranean bacteria and other forms of life feeding on the planet's chemical and thermal energy (and equally possible on, or rather *in*, other planets too, by the way) actually outweigh all life on the surface.[9] Our own bodies harbor billions of bacteria, while we ourselves may play a role like our bodies' bacteria in the larger living organism which is now supposed, by some scientists, to include all life on earth.

These are the simplest of things: bacteria, soil, rock, air. The air itself is a kind of ocean in which we all live, a world-swaddling sea that weighs, altogether, 5,000 trillion tons. In truth we live *in* the Earth, not *on* it. In some ways the atmosphere itself is the most dynamic ocean of all, never still, full of gasses and dusts and spores and fungi; and animals of all sorts, bats and gnats and eagles, flutter-bys and bees; pollen and leaves and seeds. Air courses through our own bodies constantly, 450 cubic feet a day, 10,000 or so breaths. We are constantly and unavoidably intimate with the world; not to be intimate in this way means quick suffocation, death.

There are wild things (or: Wild Things, following that great philosopher, Maurice Sendak) right in our kitchens. Cockroaches, for example, astonishing creatures, sprinting the human equivalent of 200 miles an hour on split-second notice, so preternaturally sensitive to vibration that they are used by scientists in touch research. If cockroaches were endangered, we would regard them as one of nature's greatest marvels. (Not to worry, though: they'll outlast us.) A species of leaf-eating weevil, 1 inch long, camoflauges itself by carrying a forest of tiny ferns and mosses in crevasses on its back. Still tinier insects live in that forest. Worlds within worlds. A Daddy Longlegs periodically takes up residence in my car, rides back and forth with me to work. I see one of his legs sticking out from underneath the glove compartment as I drive. Computers have "bugs" worldwide, and the original "bug" was, of course, a bug: a beetle or something that shorted out the first UNIVAC. Long live bugs!

How far we might come from the TV world of crisis, from an Earth lost in abstraction and talking heads, one more thing to learn about in school! I once took a PhD seminar (it's never too late) in the philosophy of nature to a wigwam in Hauppauge, Long Island, New York. Half a mile from the Nothern State Parkway, but we did not hear the cars, perhaps because we were so immensely distant from the Parkway and all it stands for in time if not in space. It was drizzling, foggy, and cold. The mists from our voices rose to join the smoke seeking the small smoke-hole. We passed a cup, alerted to the calls of geese and wild turkeys—penned, to be sure, but still calling excitedly in the fog. We sat on the ground and talked about Abenaki creation stories. We spoke slowly, for once, not covering all other sounds with our voices. And that small matter of tempo in the end may be the best mirror of the intuition the stories spoke of, that sense of being simply part of a larger living world. There was something to hear besides ourselves: the calls of the birds and the wind; our fire echoing the hiss of the rain; and our voices, when we spoke, interweaving among the animals'. *This* is what it means to feel, as Native Americans always said, that "The Earth is Alive"—and that we are part

of it. And it is all there, all still there, even on Long Island. Denise
Levertov writes in "In the Woods":

> Everything is threatened, but meanwhile
> everything presents itself:
> the trees, that day and night
> steadily stand there, amassing
> lifetimes and moss, the bushes
> eager with buds sharp as green
> pencil points. Bark of cedar,
> brown braids, bark of fir, deep-creviced,
> winter sunlight favoring
> here a sapling, there an ancient snag,
> ferns, lichen. And the lake
> always ready to change its skin
> to match the sky's least inflection . . .[10]

It is not and never can be "too late" to "save" *this* Earth, the
Earth of breeze and tide, fog and bugs, lichen and moss, subterranean bacteria all a-churning and the continents themselves with
all their indomitable ranges sliding around on molten rock, up-
setting all the balances and changing all the maps. Perhaps this is
the final and most decisive reduction introduced by TV: it reduces
nature to what we have lost, or are losing. And the losses are certainly many. But they are not the whole story—not at all.

Change Begins at Home

So maybe this Earth does not need *our* saving, anyway, as if once
again humans must come to the rescue in the last act, using that
distinctive big brain of ours to put things right. It turns out that
we do not even have the biggest brain around: that honor goes to
whales, some of whom have brains up to six times bigger, with
whole regions, masses of cortex, that we don't begin to understand.
So just maybe, if we really are living out some sort of global
drama, we are not the heroes, despite all our space travel and blaz-

ing cities and everything else. Maybe we are just extras. On the "Gaia Hypothesis," the theory that the Earth is in some sense a single living organism mediated through the atmosphere, it has been alleged that the chief human contributions are the gasses produced in our intestines, and even at outgassing we are, of course, bested hands-down by the ruminant herbivores.

A more suitable and circumspect role would be to look to ourselves. Who *we* really need to "save" is *us*. I don't mean simply looking to our own survival, though that wouldn't be a bad idea either. It is our task, now, coming into some sort of ecological awareness, to learn to live in accord with that awareness, to *learn to live as co-inhabitants of this planet*. We need to learn what might best be called a certain kind of *etiquette*. And this takes work—but on the whole it is not the kind of work we have yet had in mind when we think of responding to the environmental crisis.

We must stop what destruction we can stop—of course. Recycling and all the rest—of course. But the work that is harder to recognize are the changes necessary in more personal, everyday patterns of attention. Watch the spiders. Watch the skies. Walk. Garden. Let the lawn go wild. Feed the birds. *Learn* the birds. Talk to the animals. Seek out the stories of your place, pay attention to the names. This is the "etiquette" I mean: not claiming all the space for ourselves, learning to listen, learning how to *invite* the larger world, other presences, to re-enter our lives.

Suppose that certain places were set aside as quiet zones: deliberately protected areas, where cars, lawnmowers, stereos, and their kin do not define the soundscape, therefore a life shared with the other-than-human in the simplest ways: winds, birds, silence. We might quite literally "come back to our senses." If bright outside lights were also disallowed, we could see the stars at night, see the moons wax and wane, and feel the slow pulsations of the light over the seasons. The heaviness of the night could return. The stars could return, and the night creatures now exiled by the light.

This is not a utopian proposal. Unplug a few outdoor lights, reroute some roads, and in some places of the country we have a first approximation, even when the electricity is on. Even in my *city*, it is still possible to sit out on the back porch with my daughter in my arms and rock her to sleep with the owls and the stars. And what it would take to preserve and extend such spaces, in many regions and corners, is not necessarily so great. Return more neighborhood roads to local traffic only. Preserve owl habitat, plant wildflowers. Instead of more and more tract developments consuming cornfields and woodlots, let us try some experiments in creative zoning, make space for increasingly divergent styles of living on and with the land: experiments in recycling and energy self-sufficiency, for example, or mixed communities of humans and other species. Or other possibilities not yet even imagined. Canadian ecophilosopher Alan Drengson proposes the creation of "ecosteries"—"centers, facilities, stewarded land, nature sanctuaries, where ecosophy [ecological philosophy] is learned, taught, and practiced"—on analogy to the medieval monastaries: "places where spiritual discipline and practice are the central purpose." There is no reason that we must condemn ourselves to another ten thousand suburbs all the same.[11]

In the midst of the worst city we can still imagine little "pocket parks," strategically placed, insulated from noise. "Quiet backs" are common in the older cities of Europe—small green areas, behind houses or public buildings, densely planted, perhaps connected by small footpaths and waterways—like the walk through the cathedral close in Chichester cited by Christopher Alexander and his colleagues in their synoptic tract *A Pattern Language*, where, "less than a block from the major crossroads of the town, you can hear the bees buzzing."[12] This is not only in the city but in the very middle of the busiest part of the city. The remergence of the more-than-human even in the city is not at all impossible, but we must *plan* for it.

Alexander and his colleagues aim to spell out the patterns, often ancient though not necessarily even fully conscious, that define the most livable and fulfilling of our cities, neighborhoods, and houses. They propose interlocking "city-country fingers" that bring the open countryside within a short walk or bicycle ride from downtown. They calculate the maxiumum distance from home that a pocket park can still attract walkers (2–3 blocks). They calculate the optimum size for such parks. They uncover the patterns that underlie the attraction even of small but "enchanted" natural places, again in the very midst of the city: "layered" (gradual, phased) access, the presence of running and still water, the presence of animals (birds, snakes, goats, rabbits, wild cats). They plead for "site repair," for building on the *worst* parts of a piece of land rather than the best, so as to repair and improve the poorer parts while preserving the most precious, beautiful, and healthy parts (and honoring the fact that these parts are often slowly evolved and complex, not something that can be recreated elsewhere even if we or the "landscape contractor" try). They argue for the necessity of what they call "positive outdoor space": places partly enclosed by buildings and natural features so as to have a shape of their own, courtyards or partial courtyards, for example, as opposed to the shapeless outdoor space so familiar around the squarish and irregularly placed buildings of our suburbs and cities, and for "half-hidden gardens": neither the entirely decorative traditional American front yard nor the wholly private back gardens of Europe, but an intermediate kind of space.

How different life could be! Notice that we are not talking about sweeping, dictatorial, disaster-driven social change, the sort of thing that the word "environmentalism" usually implies on the news and in our politics. No: here we speak of tinkering with zoning requirements, building or retro-fitting our cities and neighborhoods in small ways that take time, like the shrubs that may take a generation to grow up to create half-hidden gardens. Rethinking the house. Modern American houses all too often function as fortresses against the supposed dangers, human and non-human, of the "outside" world. It becomes hard for us to imagine anything else. Yet here too there are alternatives, in fact entire alternative traditions. Frank Lloyd Wright used the wall, freed from its support functions, as a delicate and deliberately ambiguous transition-point between outside and in. Traditional (pre-air conditioning) Southern houses half-buried the first floor for coolness and used breezeways to amplify the faintest breeze—the winds were invited in, like friends. Native American styles still dominate parts of the Southwest, like adobe, made from the very clay of the building site, periodically replastered with the same. The buildings literally grow out of the earth. Could we not make a practice of recovering native and traditional architecture, wherever we happen to live?

No doubt bees in the parks and adobe houses are not quite what one expects of modern-day environmentalism. It is easier to bemoan the lost wilderness than to teach your children the constellations in your backyard. But even Thoreau at his cabin on Walden Pond, who it gratifies us to think of as such a hermit (so that we can also say: his life is no longer possible for *us*), in fact lived within a mile of Concord and walked there nearly every day to see his family. He lived close enough to the road that he could smell the smoke of passing pipesmokers. Fellow townspeople and farmers fished in the pond; the railroad went by one edge of it. The classic meditation on the human relation to nature, written from virtually within the city limits of Concord? But how appropriate! The "rest of the world" is not somewhere else, but *right here*. Correspondingly, the way back might be a little different than we think too.

Transhuman Etiquettes

Consider finally and again very briefly our relations to our fellow creatures, other animals. Not, by the way, just "animals," as if we weren't animals too: that little piece of language is already a first and fundamental point of etiquette here. We too are animals.

We have still not managed the most elementary politeness with respect to other creatures. Ten years or so of research aimed at getting captive apes to *talk* dead-ended when it finally dawned on someone that they don't have our kind of vocal equipment. Now we try to get them to manipulate symbols, or use sign language (ASL), and the success of some animals has been stunning. Still, why should we suppose that they even *care* about using symbols? At least outside captivity, already itself a stunning refusal of even the most minimal etiquette. Chimps trained in ASL stand by the doors of their cages begging for the keys. "You stay, I go," they say. Now that funding has dried up (Eugene Linden speculates that the research went a little *too well*) they are often caged with keepers who do not know ASL at all.[13]

For starters, surely, we need to ask how other creatures might care to live with *us* (or not), rather than taking it upon ourselves to simply define and thereby limit them too. It is not just that the symbol-using research on apes, for example, is a little ambiguous. It is better to ask: on what terms and by what means would an ape—a free ape, not a captive one—care to communicate with us at all? And apes are our very nearest relatives. What could be the "terms" of, say, a dolphin, utterly at home in the waters, who in all probability can "see" inside his/her companions by echolocation? What would language even mean for such a creature?

So it may be that we really know almost nothing about the real possibilities of dolphins, apes, and most other creatures too: perhaps very little even about our *own* possibilities! Once again, *we* are the ones who really need "saving." We are the ones who need to approach *them* in a different spirit.

Jim Nollman plays jazz rhythms with killer whales, orcas. Working in their media, as it were, not primarily ours, and going to them in a way that allows them to break off the encounter whenever they wish. Talk about elementary politeness: this is the most basic considerateness of all: not forcing your presence or your projects on another. Millions of official research dollars pour into academic research on captive animals—or dead animals, the perfect subjects—while Nollman just jury-rigs a floating drum or puts his guitar in the canoe and paddles out to visit his friends.

To announce himself, he plays chords through an underwater sound system. Sometimes the orca come, sometimes they do not. He recounts some of the nights when they came:

> [One night] it seemed as if the whales vocalized constantly, not at all coordinated with the harmonic and rhythmical structure of the chord progression. [But] on the second night . . . one individual whale stepped out to take a kind of lead voice with the guitar playing. The rest of the pod chose to stay in the background, jibber-jabbering among themselves in a quieter tone which seemed unrelated to the unfolding ensemble playing at

center stage. At the same time that the whales split into singer and Greek chorus, a group of humans appeared at the seaside sound studio. . . . They, too, began to comment among themselves at key places in the interaction. Sometimes the human observers would comment at the same moment that the observing orcas seemed also to comment. Once, the correlation was so clear that I had to stop playing a moment, just to get my bearings. . . .

The third night evolved into pure magic . . . I began . . . by mimicking the standard stereotypical vocalization of the pod: a three-note frequency-modulated phrase that begins and ends on the D note. But this pattern is never frozen. Rather it varies in form by the addition or deletion of the speed of the glissando, by the fluidity of the legato. In other words, the whales' own language varies just exactly the same way that a jazz musician varies a standardized melody. And the whales seemed very aware of my own attempts to vary their own song by ending each of my phrases with a solid obbligato amen of D to C to E to D.

Unfortunately, the highest note available to my electric guitar is a mere C-sharp, an impenetrable half-step universe below the orcas' tonic note. Thus, in order to reach their register, I needed to bend the high string—something ordinarily not that difficult—but, in fact, rather clumsy to achieve hunched up in the dark fog while fingering up at the very top of the guitar neck. The first time I attempted the bend, the result sounded like a very respectable approximation of the orcas' own phrasing . . . [I] repeated the phrase a second time. Suddenly, the high E string snapped. While I sat there in the thick night air fumbling through my guitar case for a fresh string, the orcas stepped up the intensity of their vocalizations. Calling, calling for me to rejoin the music. Every so often one of them would punctuate a long sinuous phrase with the obbligato.

I tightened up on the E string, and stubbornly plucked out the orcas' obbligato, but this time in C-sharp instead of D. The centerstage orca immediately answered by repeating the phrase in C-sharp. Otherwise, it was the exact same melody. From that point on, the dialogue between us centered around the common C-sharp chromatic scale. And the conversation continued for more than another hour in very similar fashion. . . . What the orca and the guitar player settled upon was the conversational form of dialogue. Each of us waited until the other had finished vocalizing before the other one started. In order for such a form to work properly, both of us had to become acutely conscious of each other's beginnings and endings. Once and a while, one of us would step out before the other one had completed his piece, but in general, the form of the dialogue was clearly working. And as such, the resultant musical exchange never digressed to a mere call and response. . . . There was always a feeling of care and of sensitivity, of conscious musical evolution within the time frame of a single evening's music. I might play three notes and the orca might repeat the same progression back to me, but with two or three new notes added on the end. Once, I made an error in my repetition of one of the orca's phrases. The whale repeated the phrase back again—but this time at half the speed!

After an hour of this intense concentration . . . [t]here was nothing else to do, no place else to go with the dialogue but directly into the sharply etched reggae rhythm of the previous two nights. I played it, inexplicably, in the key of A. The orca immediately responded with a short arpeggio of the A chord. When I hit the D triad on the fifth downbeat, the orca vocalized a G note, also right on the fifth downbeat. It was the suspended note of the D triad. Then back to A and the orca responded in A, again on the downbeat. The agile precision of rhythm, pitch, and harmony continued through the entire twelve-bar verse.[14]

Jamming with whales! And we think we know what is possible in this world, enough to say that all is lost?

> Utilizing the language of my own musical training, it feels very comfortable to name such an encounter a jam session. . . . But perhaps I stand guilty of bald-faced anthropomorphosing [sic]. In other words, for [the orcas'] signature whistles to be called music, must not the orca hold a concept that is at least analogous to what we humans know as music? I disagree. What we invented was neither human nor orca. Rather, it was *interspecies* music. A co-created original.[15]

Nollman is not Saving the Earth. He is not even saving orcas. He is *joining* the orcas, with grace, with skill, with etiquette. Free and wild. Enough to hope for just that.

Across all the species and across every expressive medium there may be similar possibilities. The whole world sings. There are birds singing right outside your window, right now. Sing back. In all seriousness: sing back. On a bird walk with my environmental ethics class, several years ago, one student started whistling with a mockingbird. A ten-minute dialogue ensued; the rest of the class just stood there, agape. You can talk with a bird?

Birds are interesting as one form of more-than-human intelligence that is almost always present with us outside, though seldom attended to. Climbing in the hills or walking in rolling country, you learn to watch for unusual circling or gathering over the next hill. The birds tell you things you wouldn't otherwise know: something has died, something threatens, something has happened over there. In this way we begin to recognize a kind of sensory co-presence in the land, not at all so exotic as orcas, but, as I say, virtually omnipresent. As I drive the freeway I see the hawks perched on the high-tension wires at particular spots, or circling along with the turkey vultures overhead. Even here there is a sense of an animate presence beyond the human that broadens and deepens the human world. And correspondingly *they* watch *us*. Condors, for instance, are curious birds, like most scavengers: they enjoy watching us, apparently, which is one cause of their high mortality rate.

Other words come to mind in speaking of etiquette: tact, courtesy, generosity, humility. An alligator scientist jumps in an alligator pond (with a stick): "I was pretty sure the alligators were communicating with subtle visual signals. Slight changes in body posture and body elevation in the water—things like that. But being on this boardwalk looking down on them, I wasn't able to see those slight changes very well. I thought that if I got at an alligator's eye level, it would be pretty easy for me at least to see what's relevant to an alligator."[16] A wonderful image. *Looking down* on the animals, he couldn't begin to understand them. Joining them, he could—or could, at least, "see what's relevant to an alligator."

It turns out that alligators, as well as other "armored" animals like turtles, are extremely sensitive to touch—odd as it may seem: in fact alligator courtship is mostly a matter of, literally, "neck-

the spectators to join the dance. Not just anyone. One dancer has fascinated us all night; a great burly character, festooned, madcap but still weighty, bear-like. Now over he comes, straight to my daughter, hand extended, no words. My child, for half of her short life fascinated with bears, fearful and intrigued at the same time, beckoned by a bear. Off she goes. My world spins away from me.

When he dances her back, we talk. He's North Dakotan, he tells us his English name and his work. Then his Indian name. Dancing Bear. Indeed, a dancing bear. My child, bear-entranced—how could he possibly have known this?—beckoned by a bear.

It is "too late"? I have only to watch my six year old Bear Dancing to know better. Here in this crystalline moment is signalled the possibility that we—we descendants of those who came here from elsewhere, but born here ourselves—might finally begin to inhabit this land as *natives*: that is, to take up our nativity, the fact that we were after all born here, as a challenge and an invitation. We may yet *become* native Americans. Not in the sense that we will or even could somehow imitate those who we now call Native Americans: that would be only to join the "Wannabe" tribe, as they call it in just derision. No: we (all of us, now, including those who the Canadians more aptly call First Nations peoples) must find our own way, together, a new way no doubt, not a way that denies the past or the manifold traditions that the emigrants brought to this land, but no longer insensitive to the land either. The writer Barry Lopez speaks of this as "*re*discovering North America." We may well be inspired by the First Nations peoples, the original native Americans, who after all did co-inhabit this continent, relatively peacefully, for ten or twenty thousand years (they would say, forever), with that profusion of life that so amazed the first Europeans, and that was then so quickly dispatched. In the end, though, it may be that the chief inspiration must come from the land itself, and its creatures, as it did and does for First Nations peoples themselves. But it is coming. I see it as I watch, through my tears, my German-Russian-English-Jewish child dancing with a bear. Coming home. How could it be too late? In the moving unbalanced balance of things, the Earth saves us just as much as we "save" the Earth.

Notes

1. This and the opening paragraph draw upon Farley Mowat, *Sea of Slaughter* (Boston: Atlantic Monthly Press, 1984), especially pp. 135, 139, 141, 218, and 294–95, and on Paul and Anne Ehrlich, *Extinction* (NY: Ballantine, 1981), p. 137.

2. On current extinction rates, see William Stevens, "Botanists Contrive Comebacks for Threatened Plants," *New York Times*, 11 May 1993, p. C1. On blue whales, see Peter Dobra, "Cetaceans: A Litany of Cain," in Donald Vandeveer and Christine Pierce, *People, Penguins, and Plastic Trees* (Belmont, CA: Wadsworth Publishing Company, 1986), p. 130.

3. On wolf subspecies, see Barry Lopez, *Of Wolves and Men* (New York: Scribner's, 1978), pp. 14–16. On bats, see Dianne Ackerman, *The Moon by Whalelight* (New York: Random House, 1991), pp. 46–47. On blue and right whales, see Dobra, "Cetaceans: A Litany of Cain," pp. 127, 130–31. I have seen slightly higher numbers recently in the popular press.

4. Susanna Hecht and Alexander Cockburn, *The Fate of the Forest* (New York: Verso, 1989); Mowat, *Sea of Slaughter*, pp. 139–40; Tom Jay, "The Salmon of the Heart," in Finn Wilcox and Jeremiah Gorsline, eds., *Working the Woods, Working the Sea* (Port Townsend, WA: Empty Bowl Press, 1986), p. 111n.

5. Kirkpatrick Sale, "The US Green Movement Today," *The Nation* 19 July 1993, p. 92.

6. On the points in this paragraph, see Christopher Stone, *The Gnat is Older Than Man* (Princeton: Princeton University Press, 1993), pp. 20–23. For the newest data on global warming, and some discussion of the intepretation of satellite data, see Odil Tunali, "Global Temperature Sets New Record," in Lester Brown, et al, *Vital Signs 1996* (Washington, D.C.: Worldwatch Institute, 1996), pp. 66–67.

7. On undiscovered dinosaurs, see Michael Lemonick, "Rewriting the Book on Dinosaurs," *Time*, 26 April 1993, p. 43. On the coelacanth, see Keith Thomson, *Living Fossil: The Story of the Coelacanth* (New York: Norton, 1991).

8. Natalie Angier, "Animals and Fungi: Evolutionary Tie?," *Raleigh News and Observer* (16 April 1993), p. 14A.

9. Thomas Gold, "The Deep, Hot Biosphere," *Proceedings of the National Academy of Sciences* 89 (1992), pp. 6045–49.

10. Denise Levertov, "In the Woods," in W. Scott Olsen and Scott Cairns, eds, *The Sacred Place* (Salt Lake City: University of Utah Press, 1996), p. 206.

11. Alan Drengson, "The Ecostery Foundation of Noth America: Statement of Philosophy," *The Trumpeter* 7 (1990), pp. 12–26.

12. Christopher Alexander, *A Pattern Language* (New York: Oxford University Press, 1977), p. 303.

13. Eugene Linden, *Silent Partners: The Legacy of the Ape Language Experiments* (New York: Ballantine, 1986).

14. Jim Nollman, "What Seagull Says to the Orca," in *Dolphin Dreamtime* (New York: Bantam, 1987), pp. 146–50.

15. Ibid., p. 148.

16. Quoted in Dianne Ackerman, *The Moon by Whalelight*, pp. 64–65.

3

———•———

Paths Beyond Human-Centeredness

Lessons from Liberation Struggles

Val Plumwood

1. Introduction: Seeing Nature Differently

Many environmental philosophers have said that western culture is ecologically destructive in its dominant forms primarily because it is human-centered, or "anthropocentric." A human-centered and self-enclosed outlook deeply rooted in our culture has, they suggest, caused us to lose touch with ourselves as natural beings, embedded in the biosphere, and with the same dependence on a healthy biosphere as other forms of life. We have developed conceptions of human identity as belonging to a sphere apart, outside of and above nature and ecology—a sphere of ethics, technology, and culture. And so we have also tended to assume that ethics and value are exclusively concerned with and derived from the human sphere. We do not need ethical philosophies to guide our relationships with the natural world into ethically and ecologically sound forms, since nature is just a resource we can make use of however we wish.

Philosophers are divided about the need for a challenge to this dominant framework of Western thought, as well as about how to go about such a challenge and how far ethical extension to the non-human world should go. Some philosophers have argued that human-centeredness is inevitable. There is no way, they say, we can ever throw off our human conceptual apparatus to see the world differently. Some also complain that the philosophical critique of human-centeredness provides little help with practical environmental activism, or with strategies and goals for a movement to stop ecological damage. Work for change is best undertaken within the dominant framework of thought, they argue, but with more emphasis on science and on future people.

Other philosophers, less strongly wedded to tradition, respond that abandoning the challenge to human-centeredness risks a loss of courage and vision that would reduce environmental philosophy to a cynical game where the players accept the rules of a deeply problematic framework because it is too hard for them to imagine a way to think differently. This is just the sort of place, they think, where philosophy could be most useful, helping us to imagine and formulate alternative ways to think about ourselves and nature and to restructure our lives. And so the debate goes on.

I shall argue that a look at other liberation struggles can help us here. Critiques of "centrism" are at the heart of modern liberation politics and theory. Feminism has focussed on male-centered-ness, also called "androcentrism." Anti-racist theory has focussed on ethnocentrism and eurocentrism, which takes a particular culture or ethnicity (in the case of eurocentrism that of Europe) to be central. Gay activists have criticized hetero-centrism, which treats heterosexual relations as the basic model or "center" in terms of which others are deviant, and so on. The green movement's flag-ship in this critical armada has been the critique of human-centeredness. Surely the critique of this form of centrism could learn from some of these relatively successful (at least, relatively well-formulated) others. In fact, understanding the struggle against human-centeredness in the context of the struggle against other kinds of centrism might be exactly what is needed to overcome some of the problems critics have seen in it.

2. Is Human-Centeredness Inevitable?: The Question of Prudence

If there is anything that critics of human-centeredness agree about, it is that we must avoid treating nature as if it were merely a means to our ends. However, it is sometimes argued in response that an ethics based on human interests is not only all that is needed for the conservation of nature but all that is *conceivable*.[1] We are humans; we cannot avoid thinking in terms of our own interests. In fact, if somehow we actually could put our own interests completely aside, we would be left with a totally useless ethics. No one would find it compelling. An ethics that considered only effects on nature and ignored humans would be irrelevant to the practical politics of environmental activism and would cut itself off from real policy debates.[2]

This objection certainly has to be taken seriously. We do need, as humans, to take good care of ourselves, not leaving ourselves unsafe, unprotected or unprovided for. This is what philosophers and moralists call "prudence." "Prudential" arguments in this context, then, would be arguments for avoiding certain environmental practices which consider the effect of those practices on the safety, survival, and welfare of human beings. Ozone depletion and pollution harm human health, overfishing destroys resources for future humans, global warming could unleash potentially catastrophic climatic change and extremes, and so on. If the core theoretical distinctions of environmental philosophy indeed tell us that it is human-centered to take good care of human interests, if they force us to condemn as human-centered all criticisms of our treatment of nature that refer to the damage its degradation does to human beings, then they would indeed make the ideal of escaping human-centeredness quite impractical. And if, as some critics go on to argue, the ideal of avoiding human-centeredness also

ı

provides only vague alternative reasons for avoiding environmentally degrading actions, it is a real liability for practical action.

But are we in fact forced to condemn as human-centered all prudential types of environmental argument? I think this is a misinterpretation of human-centeredness as well as a misinterpretation of prudence. Consider the parallel case of egocentrism. We would usually say that someone was egocentric if, among other things, that person consulted only their own outcomes, welfare, or interests in deciding what courses of action to follow, and ignored outcomes for others or failed to consider them as good reasons for or against the action being considered (this is the extreme case; often we would say someone was egocentric when they just gave other people's interests excessively low weight). But the definition of prudence as taking care for and protecting yourself does not imply that you cannot *also* take care of others, any more than your taking care of orange trees means that you cannot *also* take care of lemon trees. Considering your own interests does not imply that you cannot also consider others' interests as well as, or as related to, your own. The idea of prudence says nothing about consulting your own interests *to the exclusion of others*. That is not prudence, it is selfishness or egocentrism.

Similarly, the ideal of avoiding human-centeredness does not imply at all that humans should not be prudent, or that we cannot consider the effect of environmental damage on our own human interests along with the effect of our actions on other species and on nature generally. The critics' objection rests on identifying prudence with something much stronger—with a kind of species selfishness that treats other beings *solely* as means to our own, human ends. The philosopher Immanuel Kant tells us that humans are to be conceived as ends-in-themselves and cannot be treated as merely means to our ends, and though Kant himself restricted this kind of standing to humans, environmental philosophy typically proposes to generalize it. But the crucial phrase here is *"no more than."* We must inevitably treat the natural world to some degree as a means, for example as a means to food, shelter, and other materials we need in order to survive, just as we must treat other people to some degree as means. In the circus the performers may make use of one another by standing on one anothers' shoulders, for example, as a means of reaching the trapeze, but our obligation to avoid using others solely as means (or *"instrumentalizing"* them, as philosophers term it) does not imply banning the circus. What is prohibited is unconstrained or total use of others as means, reducing others to means—tying some of the performers up permanently, for example, to use as steps.

In short, then, prudential reasons and non-prudential reasons for action are not mutually exclusive. Prudential and non-prudential reasons can combine and reinforce one another, and may not always be sharply separate, since any normal situation of choice always involves a mixture. The problem lies rather in the refusal to go beyond questions of human well-being. Nonhumans are wholly excluded from morality and are treated as mere tools, unworthy of moral consideration in their own right. Only by identifying prudence with this *radical* kind of selfishness can critics discover a malaise in environmental ethics. To be prudent in our dealings with nature is both essential and benign from the perspective both of nature and of ourselves; while to be governed by egocentrism or by instrumentalism in our dealings with nature is damaging but far from inevitable.

3. Is Human-Centeredness Inevitable?: The Argument from Standpoint

The argument just considered concludes that anthropocentrism is unavoidable because some degree of prudence is unavoidable. A related argument reaches the same conclusion by arguing that some kind of human *standpoint* is unavoidable. This kind of argument for inevitability has appeared frequently over the years of debate in environmental ethics, but I will look at the most recent and perhaps also the most uncompromising rejection of the ideal of non-anthropocentrism, an argument recently proposed by William Grey.[3]

In Grey's "cosmic" sense of anthropocentrism, a judgment can be claimed to be anthropocentric if it can be made to reveal any evidence of dependency on a human *location* in the cosmos, on human scale or "human values, interests, and preferences."[4] Grey motivates his cosmic reading of this key concept of environmental philosophy by analogy to the shift from a geocentric (earth-centered) to a heliocentric (sun-centered) view of the universe. Prior to the Copernican revolution, the sun was assumed to revolve around the earth, but Copernicus defeated this geocentrism by taking a more impartial and modern view of the universe which showed the earth to revolve around the sun. In the same way that Copernicus overcame "parochialism" by moving to a less limited, less earth-centered viewpoint, so overcoming anthropocentrism, Grey thinks, requires a move away from human locality and human perspective to a view of the world *"sub specie aeternitatis,"* through cosmic rather than human spectacles. Thus, according to Grey's account, to defeat anthropocentrism we must distribute our preferences with perfect detachment and perfect impartiality of concern across humans and non-humans, to achieve at last a view-from-nowhere which abandons all specifically human viewpoint on or preference about the world. It is no surprise that this turns out to be impossible, and Grey then proceeds to the conclusion that anthropocentrism is vindicated and environmental philosophy in general shown to be misguided.

As Grey sets them out, then, the basic steps in the argument for the inevitability of cosmic anthropocentrism are these:

Premise 1. To avoid anthropocentrism, we must avoid any reliance on human location or "bearings" in the world, any taint of "human interest, perception, values or preferences," "human standards of appropriateness," "human concerns," and such telltale signs of human origin as "recognizably human scale" and favor for the human species.[5]

Premise 2. But this task is impossible, as demonstrated by various examples of absurd results, the discovery of obvious human reference in normal judgments, and by general argument. For example, Grey argues that we cannot adopt a completely impartial time scale without losing "recognizably human scale," and that when we lose this, there is no ground for preferring any one state of the universe over any other.

Conclusion. Therefore anthropocentrism is unavoidable, and the demand to avoid it is conceptually confused.[6]

I think it can be conceded that it is impossible for humans to avoid a certain kind of human epistemic locatedness. Human knowledge is inevitably rooted in human experience of the world, and humans experience the world differently from other species. Nevertheless, this kind of human epistemic locatedness is not the same as anthropocentrism, adequately understood. Grey's argument runs together two very different things: ethical interest and epistemological locatedness. In order to treat another person or being with sensitivity, sympathy, and consideration for their welfare, we may often need a process of ethical reflection. This means, according to Iris Young, that we may have to take some distance from our own immediate impulses, desires, and interests in order to consider their relation to the demands of others, their consequences if acted upon, and so on. But as Young notes, this process of standing back a little from the self does not require that we adopt a "cosmic" point of view emptied of *all* particularity and all trace of our own location, that is somehow universal or the same for everyone, and it is hard to see how such a "view from nowhere" could lead to any action at all.[7]

Moral reasoning requires some version of empathy, putting ourselves in the other's place, seeing the world to some degree from the perspective of an other with needs and experiences both similar to and different from our own. This may be said to involve some form of enlargement of or going beyond our own location and interests, but it does not require us to *eliminate* either our own interest or our own locatedness, rooting out any trace of our own experience and any concern for our own needs. If it did require this, the practice of moral consideration for others *would* be just as impossible as Grey claims the avoidance of anthropocentrism is. If we eliminated all knowledge of our own experience of suffering, for example, not only would we be unable to consider ourselves properly, but we would have no basis for sympathy with another's suffering.

The confusion involved in Grey's argument comes from a similar confusion between overcoming a narrow restriction of ethical concern to the self and eliminating all epistemic trace of self—between selfishness on the one hand and having a particular standpoint or epistemic location (locatedness) on the other. We can see from everyday experience that ethical concern should not be identified with mere location, for the fact that we are located, partially at least, in our own experience or that of our cultural group does not mean that we cannot and should not be considerate of people other than ourselves or of people outside that group. Indeed, the same kind of argument as Grey advances would show that we must all inevitably be selfish, since we are all "located" in terms of a perspective that arises in part from our own personal experience. Once we make this distinction between ethical concern and mere location, we can begin to see why anthropocentrism is incorrectly identified as a necessary feature of locatedness and particularity. As we shall see, it is far better identified as a moral and political failing closely allied to selfishness.

There is another revealing fault in Grey's argument. Grey also "essentializes" or "reifies" the category of the human, privileging it over all other possible descriptions of ourselves we might adopt. After all, we need not have described ourselves by the term "human," since being human is only one of the things we are, and the category human is also included in many other categories. We are not only humans, but also primates and vertebrates, for example. If we use any of these other more inclusive descriptions of ourselves, we will not be able to reach Grey's conclusions that we cannot ethically consider non-human others who are primates or vertebrates. If we use any less inclusive ones, such as "Greek," "male" or "white," Grey's reasoning will oblige us to leave some categories of the human outside ethical consideration and enable us to reach some obviously objectionable conclusions, such as that male humans cannot ethically consider female humans, for example. So Grey's argument has to take the description "human" and its defining contrast against the nonhuman world to be in some sense more fundamental than any of these other possible descriptions. But there is no obvious logical case for doing this, and doing so seems to reveal a bias already in favor of the human—an anthropocentric bias, it would seem. So Grey's line of argument begs the question—it already assumes precisely what is at issue, the privileging of the category "human" and thus the validity of anthropocentrism.

There is also what we could call an "interest" version as well as a standpoint version of this same cosmic argument. The interest version argues that we must be anthropocentric because cannot totally eliminate our own human interests. But our response is much the same: why should we need to? The equivalent of Grey's assumption in the human case would be to assume that unless we can totally eliminate all concern for ourselves or our own interests, we have no alternative but to behave selfishly towards others. But we know that it is possible to consider both our own interests *and* the interests of others. Why should we have different standards for the non-human case? To aim for the total elimination of our own interest in ethical action is unrealistic and unwise, presenting us with a false choice between self-abnegation or egocentrism—either totally neglecting or being totally enclosed in our own interests. And thus, once more, "cosmic" anthropocentrism is unmasked as the product of a family of arguments which rely on shifts and ambiguities like these to demonstrate some version of philosophical egocentrism, where the crucial equivocations between locatedness and restriction of ethical concern to the self often lie buried in the concept of "selfish interest." In fact, it is no more necessary for humans to be human-centered than it is for males to be male-centered, or for whites to be eurocentric or racist in their outlook. Human-centeredness is no more inescapable than any other form of centrism.

4. The Murky Waters of Egoism: Introducing Ann and Bruce

Grey's argument for cosmic anthropocentrism is in fact an argument for human selfishness, with a structure that parallels that of similar arguments for individual selfishness. It is a species version of the perennially appealing, but long-refuted doctrine of philosophical egocentrism. But as Bishop Joseph Butler demonstrated

so clearly in the Eighteenth Century,[8] and as we all know from our
own experience, egocentrism is not inevitable at all.

We can grasp the parallel between anthropocentrism and egocentrism more easily by looking at an example. Consider the following dialogue between Ann and Bruce.

ANN: I think you ought to do a bit more of the housework
around here. I'd like a chance to write some philosophy too. I
think you're really self-centered, you only think about your own
interests, you never think about my needs at all.
BRUCE: I see you're being emotional and confused again, dar-
ling. Don't you know *everybody's* self-centered? There's absolutely *no*
way to avoid it. We *all* give weight just to our own interest. That's
what you're doing too. We're all located in space and time, none of
us can eliminate our bearings in the world or distribute our concern equally over everything. We all see the world from that perspective of the self. Inevitably our own experiences, interests, values and preferences, standards of appropriateness must color and
shape our universe, underlie everything we think and do . . .

The dialogue is interrupted as Ann throws the dishmop at Bruce,
and adds some remarks suggesting he is rationalizing his own self-
centeredness. "What's all that rubbish got to do with it ! she says.
"I'm asking you to take over some more of the housework!"

"But I *can't*" says Bruce. "That's what I'm trying to explain to
you! I can't because I can only really consider *my own* interests.
Yours don't count at all, unless I *choose* to give them some weight.
I might do that if it was *in my interest*, but you'll have to show me
how it is—how you might try to *please* me better if I did, for example. But you haven't done that yet, have you ? You've just got
angry, and . . ."

The dialogue closes with some more frustrated remarks from
Ann, which I won't repeat in detail, but which are to the effect
that Bruce is a narcissistic idiot unable to consider others. Divorce
follows shortly after. It was plainly inevitable: Ann and Bruce are
not on the same wavelength at all. Ann asks for more weight to be
given to her interests. Bruce responds with a locatedness argument

to the effect that self-centeredness is inevitable, so he can't do what she asks. Bruce's ultimate hint of a concession indicates he will agree to Ann's request, if he ever does, only in expectation of something she shouldn't be required to give, that is, for the wrong sorts of reasons, out of concern for *his* interests, not out of respect and consideration for hers. This will tell on their relationship in the long run, even if they reach a temporary compromise now.

Bruce's response is totally *inappropriate* to Ann's request. Ann asks for fairness and consideration, Bruce meets her with philosophical cant about locatedness and philosophical egoism. His and Ann's positions appear to meet, but actually they do not. That is why Ann, in the same position as the critic of human-centeredness, believes she has been fobbed off. Her respondent has not caught the sense of her claim, has the wrong kind of self-centeredness in mind, perhaps because he has been badly educated, or perhaps because he is in bad faith. There is certainly some reason to suspect bad faith, since he is using his argument to refuse to do something we know perfectly well he could do, give more weight to Ann's needs.

Now in just the same way, it seems to me, the response of the cosmic anthropocentrist completely misses the point of the case against anthropocentrism, and misrepresents the sorts of demands that are being made in terms of it. Interpreting the green critic of anthropocentrism as asking for a better deal for the non-human world—a larger share, more concern, more weight, more awareness, attentiveness—gives a rather better reading of the basic thrust of green activism than the cosmic idea of viewing the world *sub specie aeternitatis.* Not that it is never helpful to try to take the cosmic perspective, to think about our humanity in the context of the immensity of the universe, for example. In some circumstances, such a perspective may have much to offer us. My point is rather that it does not connect appropriately with the kind of ecological politics or activism we are concerned with here. Certainly the cosmic perspective is not the only perspective an anti-anthropocentrist could take.

Thus it seems that the defender of cosmic anthropocentrism responds in the same perverse way as Bruce. The defender makes a point about the inevitability of human locatedness, ultimately using it to justify a human species version of the self-centeredness and philosophical egocentrism Bruce defends. The cosmic anthropocentrist is, for the purposes of the debate over how to reconstruct the culture in more ecologically sensitive terms, operating with a similarly perverse understanding of anthropocentrism. And *of course* that concept is out of kilter with ecological politics and ecological activism. But it is by no means the only alternative.

5. Toward a General Model of "Centrism"

We have been looking at models of anthropocentrism, and correspondingly of non-anthropocentrism, that I have argued are misleading and inadequate. It is time to ask what an *adequate* understanding of anthropocentrism and non-anthropocentrism might be.

Recent struggles against ethnocentrism and androcentrism can suggest ways we might rethink and sharpen the concept of anthropocentrism in turn. Take ethnocentrism, a term used for patterns of thought and practices associated with racism and colonialism. In regard to ethnocentrism, the equivalent of Grey's position would be that in order to overcome ethnocentrism we would have to abandon all cultural location or centeredness. An historically important and much discussed special case of ethnocentrism is eurocentrism, which treats Europeans or their culture as the dominant colonial center or norm in terms of which others are deviant, peripheral, or inferior. For this case, however, recent anti-racist and post–colonial scholarship argues, quite contrary to Grey, that what is actually required in overcoming eurocentrism is not abandonment of all location but the self-recovery and critical affirmation of the Others, those cultural locations or identities that are suppressed under eurocentrism, to allow the development of a "polycentric" or an "a-centered" world.[9]

A strong current of African-American thought advocates Afrocentrism, which celebrates an African cultural heritage, as a cultural home or "center." Contemporary African scholarship has distinguished between on the one hand claiming an Afrocentric epistemic home or cultural location and on the other asserting a single dominant centrism. This distinction corresponds in part to the difference between the viewpoint of "cultural appreciation" which celebrates a particular cultural consciousness or heritage on the terms of a polycentric model, and the viewpoint of "cultural deficiency" which defines other cultures as inferior or deviant in relation to one's own.[10] It is because polycentric forms of centeredness and dominating forms of centrism are different that Afrocentrism in this polycentric form does not necessarily imply the assertion of a single, new, dominating center. Polycentric emphasis on recovering cultural location is in part a political response to being subsumed within a dominant colonial center. It is precisely because this colonial model has defined non-European cultures and races in relation to itself as Other, as periphery to center, that it is now seen as necessary to embrace a form of Afrocentrism, as a positive cultural assertion "removing African people from the periphery of the European experience to restore them to their own center."[11]

Similarly, a close look at male-centeredness (androcentrism) shows us why we can't interpret escape from centrism as the escape from *location* in the way Grey's cosmic model tries to do. The equivalent of Grey's argument in the case of androcentrism would imply that to counter the male-centeredness of culture we would need to abandon all gender location or perspective (both male and female), to achieve the gender equivalent of a view-from-nowhere. But many feminists have argued[12] that "gender-blindness" or neutrality in the context of male-dominated culture will tend to be no more than appearance and will hide rather than displace the operations of the masculine model as the cultural norm. To counter the model of the dominant center, we need positive and multiple "centering." Thus feminists celebrate women's alternative ethical and knowledge styles, and attempt a self-recovery and positive identity through women's history and women-oriented projects. Again, this need not mean replacing one form of centrism (androcentrism) by another (gynocentrism). In the context of the oppression deriving from hegemonic kinds of centrism,* the claim to centeredness, to one's own center as epistemic and cultural location or identity, takes on especial importance as a way of regaining self-definition

---

*That is, centrism with a single, dominating kind of center.—Ed.

in the face of the tendency of the single dominant centre to incorporate and assimilate elements treated as periphery or as "Other." Hegemonic centrisms exhibit a structure that I will call "Othering." Nancy Hartsock has described this structure as one which "puts an omnipotent subject at the centre and constructs others as sets of negative qualities."[13] The general logical structure involved in Othering is common, Hartsock suggests, to the different forms of centrism which underlie racism, sexism, and colonialism. The overall effect of the Othering that results from centric structure is not only to justify oppression by making it seem natural, but also to make it invisible, creating a false universalism in culture in which the experiences of the dominant "center" are represented as universal, and the experiences of those subordinated in the structure are rendered "deviant" or are not visible at all. Thus, in the case of androcentrism, cultural identity and experience is represented in terms of the experience of men[14]; in the case of ethnocentrism and eurocentrism (racism), the culture's identity and experience is represented in the terms of the experience of the hegemonic "race" or ethnically privileged group.

I outline now the chief structural features of this kind of hegemonic centrism, drawing at points on features of centrism suggested by feminists.[15] As Nancy Hartsock points out, a similar structure of Othering can be extracted from theorists of colonization. I illustrate this logic of the One and the Other with examples drawn from both male-centeredness (androcentrism) and from the colonization of indigenous peoples, especially the case of Australian Aboriginal people oppressed by eurocentrism.

A centrist conceptual structure is normally erected on the foundation of a *dualism*, a division which creates a sharp break or discontinuity between two groups, the "higher" group identified as the privileged "center," or One, and the "lower" group subordinated as its Other.[16] Full-scale dualisms have the following logical characteristics of radical exclusion and stereotyping:

1. **Radical exclusion** marks the Otherized group out as both inferior and radically separate.[17] For example, the woman is set apart as having a different nature, is seen as part of a different, lower order of being lesser or lacking in reason. We meet here *hyper-sepa-*

ration, an emphatic form of separation that involves much more than just recognizing difference. Hyper-separation means defining the dominant identity emphatically against or in opposition to the subordinated identity, by exclusion of their real or supposed qualities. It is a form of differentiation that justifies domination and conquest. Thus "macho" identities emphatically deny continuity with women—minimizing qualities shared with women. Colonizers exaggerate differences (for example through emphasizing exaggerated cleanliness, "civilized" or "refined" manners, body covering, or alleged physiological differences between what are defined as separate races.[18]) They may ignore or deny relationship, conceiving the colonized as less than human. The colonized are described as "stone-age," "primitive," as "beasts of the forest," and contrasted with the civilization and reason attributed to the colonizer.

The function of this hyper-separation is to mark out the Other for separate and inferior treatment. Separate "natures" explain, justify and naturalize widely different privileges and fates between men and women, colonizer and colonized, justify assigning the Other inferior access to cultural goods, and block identification, sympathy, and tendencies to question inequalities. A sharp boundary and maximum separation of identity enable the beneficiaries of these arrangements to both justify and reassure themselves. Sharp boundaries and discontinuity in the case of the colonized are often maintained in terms of theories of racial purity and supremacy. Typically supremacist classifications reconstruct a highly diverse field which has many forms of continuity in terms of two polarized and internally homogenized "superior" and "inferior" racialized or genderized classes.[19]

2. **Homogenisation/stereotyping** occurs when differences within an Otherized group are disregarded.[20] The Other is not an individual but a member of a class stereotyped as interchangeable, replaceable, all alike, homogeneous. Thus essential female and "racial" nature is uniform and unalterable.[21] The colonized are stereotyped as "all the same" in their deficiency, and their social, cultural, religious, and personal diversity is discounted.[22] Their nature is essentially simple and knowable (unless they are devious and deceptive), not outrunning the homogenizing stereotype.[23]

Homogenization is a major feature of pejorative slang, for example in talk of "slits," "gooks," and "boongs" in the racist case, and in similar terms for women.

Features 1 and 2 function jointly to set up the typical *polarized structure* characteristic of dualism, described by Marilyn Frye as follows. "To make domination seem natural, it will help if it seems to all concerned that the two groups are very different from each other and . . . that within each group, the members are very like one another. The appearance of the naturalness of the dominance of men and subordination of women is supported by the appearance that . . . men are very like other men and very unlike women, and women are very like other women and very unlike men."[24] This polarized structure itself is often thought of as being dualism, but dualism is usually symptomatic of a larger context of Othering which also involves the further features set out below.

3. **Denial, backgrounding:** Once the Other is marked in these ways as part of a radically separate and inferior group, there is a strong motivation for the dominant group to regard and represent them as inessential. Thus the center's dependency on the Other cannot be acknowledged. In an androcentric context, the contribution of women to any collective undertaking is denied, treated as inessential or as not worth noticing. "Women's tasks," no matter how crucial to life, will be seen as background to the masculine-associated aspects of life that are considered important or significant, and will often be classified as "natural" or as involving no special skill or care. This feature enables exploitation of the denied class via expropriation of what they help to produce. Denial is often accomplished via a biased view of what is worth noticing, of what can be acknowledged and foregrounded and rewarded as "achievement," as opposed to what is relegated to the background. Thus women's traditional tasks in housework and childraising are treated as the background services that make "real" work and achievement possible, rather than as achievement or as work themselves. Thus too the colonized are denied as the unconsidered background to "civilization," the Other whose prior ownership of and agency in the land and whose dispossession and murder is never spoken or admitted. Their trace in the land is denied, and they are

represented as *inessential* as their land and their labor embodied in it is taken over as "nature."

Australian Aboriginal people, for example, were not seen as ecological agents, and their land was taken over as unoccupied, "terra nullius" (no-one's land), while the heroic agency of the white "pioneers" in "discovering," clearing and transforming the land was strongly stressed. Similarly, although Americans grow up learning that Columbus discovered America, what he really did was to *invade* it. That America existed, was, of course, no surprise to the natives. But the usual eurocentrism entirely erases them from the picture as cultural agents. Only Columbus could be a legitimate "discoverer," and he in turn opened the hitherto "empty" land to European occupation.

4. **Incorporation** (This feature is also called "conversion"[25] and "assimilation"): In androcentric culture, the woman is defined in relation to the man as central or normal, often conceived as a "lack" in relation to him. In Simone de Beauvoir's classic statement: "humanity is male and man defines woman not in herself but as relative to him; she is not regarded as an autonomous being . . . she is defined and differentiated with reference to man and not he with reference to her; she is the incidental, the inessential as opposed to the essential. He is the Subject, he is the Absolute, she is the Other."[26] His features are set up as culturally universal, inscribed as the norm of experience; as Other she is then the exception, negation or lack of the order or virtue of the male *One*. Her difference is thus represented not as diversity but as lack or deviancy, the basis of hierarchy and exclusion.

The colonized too is judged not as independent being or culture, but as an "illegitimate and refractory foil" to self,[27] as negativity or *lack* ("backward, uncivilized") in relation to the colonizer,[28] devalued as an absence of the colonizer's chief qualities, usually represented in the West as Reason. Differences are judged as deficiencies, grounds of inferiority. The order which the colonized possesses is represented as disorder or unreason. The colonized and their "disorderly" space are available for use, without limit, and the project of the colonizer is to remake the colonized and their space in the image of the colonizer's own self-space, own culture or land, which is represented as the model of reason, beauty, and order. The speech, voice, culture, and religion of the colonized is recognized or valued only to the extent that it is assimilated to that of the colonizer. Through assimilation the Other is produced as an inferior and imitative version of the One, so that this One can retain both its claim to radical difference and also its claim to represent the whole, as the "pure" form of the Other.

5. In **instrumentalism**, the Other's independent agency and value is denied, downgraded and disvalued.[29] Traditionally, the woman is conceived as "passive" and her agency is subsumed within the agency of the male who is her "protector." She is constructed as lacking independent ethical weight, being valued as a means to others' ends rather than accorded value in her own right, deriving her social worth instrumentally, from service to others, as the producer of sons, carer for parents, etc. "Woman's nature" and "woman's virtue" are defined instrumentally, in terms of being a good wife, mother, or daughter, classically in terms of "silence and good weaving," romantically in terms of being there to please. Where she is conceived as lacking any independent value or agency, she does not present any limit to intrusion (unless this limit originates in her relationship to another male)—thus her boundaries permit or invite invasion.

Similarly, the colonized Other is reduced to a means to the colonizer's ends. The colonizer, as the origin and source of "civilized values," denies and disvalues the Other's agency, social organization and independent ends, and subsumes them under his own. The extent to which indigenous people actively managed the land, for example, is denied, and they are presented as largely passive in the face of nature. In the Australian colonizer's version of history, the indigenous Other did not present any resistance to colonization, and did not fight or win any battles, although recent historians, recovering the suppressed history, have shown that Australian Aboriginal people mounted considerable resistance.[30] Since the Other is conceived in terms of inferiority and their own agency and creation of value is denied, the colonizer thinks it appropriate that he impose his own value and agency, that the colonized be made to serve the colonizer as a means to his or her ends, (for example, as servants, "hands" or "boys"). The colonized, so conceived, cannot present any moral limit to appropriation.

One last remark. The sharing of this logic of Othering between different kinds of centric oppression helps to explain the ready transfer of metaphors between them, and the reinforcement of the ideologies of "nature" which support one kind of centric oppression by drawing on the Othering logic for another.[31] Thus racial inferiorization drew strongly on assimilating racially subordinated groups to women[32] or to animals and children,[33] all regarded as closer to nature, the sphere where a powerful structure of Otherization is apparently unquestionable. Conversely, this sharing can help to explain the way liberation perspectives and insights that disrupt this logic of the One and the Other have also historically supported one another and been transferred from one area of oppression to another, for example in the nineteenth century between women's oppression and slavery, and in the mid-twentieth century from movements against racism to feminist movements. And, as we shall soon see, the historical development in our time of a critical environmental approach to the human-nature relationship has exhibited this pattern of political transfer of insights from other liberation perspectives in an especially striking way.[34]

6. Anthropocentrism as the Othering of Nature

We can now spell out a concept of "human-centrism" parallel to the concepts of ethno- and andro-centrism outlined above. In anthropocentric culture, nature and animals are constructed according to the same logic of the One and the Other, with nature as Other in relation to the human in much the same way that women are constructed as Other in relation to men, and those regarded as "colored" are constructed as Other in relation to those considered "without color," as "white."

1. **Radical exclusion:** An anthropocentric viewpoint treats nature as radically other, and humans as emphatically separated from nature and from animals. It sees nature as a hyper-separate lower order lacking continuity with the human, and stresses those features which make humans different from nature and animals, rather than those they share with them, as constitutive of a truly human identity. Anthropocentric culture endorses a view of the human as outside of and apart from a plastic, passive and "dead" nature which is conceived in mechanical terms as lacking human qualities such as mind and agency, which are seen as belonging only to the human. A strong ethical discontinuity is felt at the human species boundary. An anthropocentric culture will tend to adopt concepts of what makes a good human being which reinforce this discontinuity by devaluing those qualities of human selves and human cultures it associates with nature and animality, and thus also to associate with nature inferiorized social groups and their characteristic activities. Thus women are historically linked to "nature" as reproductive bodies, and through their supposedly greater emotionality, and indigenous people are seen as a primitive, "earlier stage" of humanity. At the same time, dominant groups are associated with the overcoming or mastery of nature, both internal and external. For all those classed as nature, as Other, identification and sympathy are blocked by these structures of Othering.

2. **Homogenisation/stereotyping:** The famous presidential remark, "You've seen one redwood, you've seen them all," invokes a racist parallel. An anthropocentric culture rarely sees nature and animals as individual centers of striving or needs, doing their best in their conditions of life. Instead nature is conceived in terms of interchangeable and replaceable units (as "resources") rather than as infinitely diverse and always in excess of knowledge and classification. Anthropocentric culture conceives nature and animals as all alike in their lack of consciousness, which is assumed to be exclusive to the human. They are viewed as machines or automata, and the range and diversity of mindlike qualities found in nature and animals is ignored. Human-supremacist models promote insensitivity to the marvelous diversity of nature, since they attend

to differences in nature only if they are likely to contribute in some obvious way to human interests, conceived as separate from nature. Homogenization leads to a serious underestimation of the complexity and irreplaceability of nature. Thus scientists assume their own genetically engineered replacements for natural species and varieties are always superior, although they have not been tested for survival over a range of conditions as rigorously as naturally evolved varieties.

These two features of human/nature dualism, radical exclusion and homogenization, work together to produce in anthropocentric culture a polarized understanding in which the human and non-human spheres correspond to two quite different substances or orders of being in the world. As modernity developed, and the human sphere and its machines began, from about the fourteenth century onwards, to destroy and supplant in a large-scale way the non-human biome, this polarization was elaborated by the philosopher Descartes in terms of a contrast between human consciousness and nonhuman clockwork, that is, between a superior sphere of mind and consciousness and an inferior mechanistic one supposedly devoid of both. In this "mechanistic" model of nature, which was applied even to animals, nature is assimilated to a mindless machine, and like a machine, has no ends of its own but can be freely redesigned to fulfill human need. The goal of science, as the human knowledge of the world, is the manipulation and control of this machine, which promises to free humankind from death and necessity here on earth.[35]

3. **Backgrounding, Denial:** Nature is represented as *inessential* and massively denied as the unconsidered background to technological society. Since anthropocentric culture sees non-human nature as a basically inessential constituent of the universe, nature's needs are systematically omitted from account and consideration in decision-making. Dependency on nature is denied, systematically, so that nature's order, resistance and survival requirements are not perceived as imposing a limit on human goals or enterprises. For example, crucial biospheric and other services provided by nature and the limits they might impose on human projects are not con-

sidered in accounting or decision-making. We only pay attention to them after disaster occurs, and then only to fix things up. Where we cannot quite forget how dependent on nature we really are, dependency appears as a source of anxiety and threat, or as a further technological problem to be overcome.

4. **Incorporation:** Rather than according nature the dignity of an independent other or presence, anthropocentric culture treats nature as Other as a refractory foil to the human. Defined in relation to the human or as an absence of the human, nature has a conceptual status that leaves it entirely dependent for its meaning on the "primary" human term. Nature and animals are judged as "lack" in relation to the human-colonizer, as negativity, devalued as an absence of qualities said to be essential for the human, such as rationality. We consider non-human animals inferior because they lack (we think) human capacities for abstract thought, but we do not consider those positive capacities many animals have that we lack, remarkable navigational capacities for example. Differences are judged as grounds of inferiority, not as welcome and intriguing signs of diversity. The intricate order of nature is perceived as disorder, as unreason, to be replaced where possible by human order in development, an assimilating project of colonization. Where the preservation of any order there might be in nature is not perceived as representing a limit, nature is available for use without restriction.

5. **Instrumentalism:** In anthropocentric culture, nature's agency and independence of ends are denied, subsumed in or remade to coincide with human interests, which are thought to be the source of all value in the world. Mechanistic worldviews especially deny nature any form of agency of its own. Since the non-human sphere is thought to have no agency of its own and to be empty of purpose, it is thought appropriate that the human colonizer impose his own purposes. Human-centered ethics views nature as possessing meaning and value only when it is is made to serve the human/colonizer as a means to his or her ends. Thus we get the moral dualism or split characteristic of modernity in which ethical considerations apply to the human sphere but not to the non-human

sphere. Since nature itself is thought to be outside the ethical sphere and to impose no moral limits on human action, we can deal with nature as we like, provided we do not injure other humans in doing so. Instrumental outlooks distort our sensitivity to and knowledge of nature, blocking humility, wonder and openness in approaching the more-than-human, and producing narrow types of understanding and classification that reduce nature to raw material for human projects.

7. Paths Beyond Anthropocentrism: Practical Implications of the Model

Understanding anthropocentrism as a form of Othering points naturally to an alternative set of strategies for moving beyond anthropocentrism, strategies that avoid the perils of imprudence and impracticality that bedevil the accounts of anthropocentrism considered in sections 2 and 3. On this model, moving beyond anthropocentrism does not require that we humans perform the impossible feat of abandoning all human "location." It does require that we abandon the human ethical and political equivalent of self-centeredness and self-enclosure. This task is not only feasible—it is actually what good ecological activism is geared to accomplish.

Ecological thinkers and activists try to counter *radical exclusion* (the first feature of anthropocentrism on the Othering model) by emphasising human *continuity* with non-human nature. By bringing about a better understanding of human embeddedness in nature, we contest *dualized conceptions* of humanity (the second feature) which treat humans as "outside nature" and above the ecological fate which has overtaken other species. We aim to challenge or disrupt conceptions of human identity and virtue based on the exclusion or devaluing of characteristics shared with non-humans, such as emotionality and embodiment. We stress instead human relatedness to and care for the natural world. We should remember also that terms like "nature" lump seals and elephants along with mountains and clouds in the one sphere of alleged mindlessness. This is a kind of internal homogenization (another aspect of the second feature of Othering), which we contest by promoting an understanding of nature's amazing diversity.

To counter *backgrounding and denial* (feature 3), ecological thinkers and green activists try to raise people's awareness of how much we all depend on nature, and of how anthropocentric culture's denial of this dependency on nature is expressed in local, regional, or global problems. There are many ways to do this. One important way, for those with a theoretical bent, is to criticize institutions and forms of rationality which fail to acknowledge and take account of human dependency on nature, such as conventional economics. Through local education, activists can stress the importance and value of nature in practical daily life, and create understandings of the fragility of ecological systems and relationships. Those prepared for long-term struggles can work to change systems of distribution, accounting, perception, and planning so that these systems acknowledge and allow for nature's needs and limits. Bringing about such systematic changes is very much what political action for ecological sustainability is all about.

There are also many ways to counter *incorporation* (the fourth feature). We can work in many cultural fields to displace the deeply rooted traditional view of non-human difference as "lack" and the devaluation of non-humans as inferior versions of the human species, and aim to replace it with an affirmation of non-human difference as an expression of the richness of earthly life, and a view of non-humans as presences to be encountered on their own terms as well as on ours. In terms of biological education, activists work to counter incorporation when they create an understanding of nature's own complex ecological order, and of the developmental story of species and of the earth. Activists of all kinds work to counter incorporation also when they engage critically with the systems and institutions which imperil the precious non-human presences around us, such as conventional economics, or when they oppose destructive development of areas which carry their own complement of more-than-human life. In opposition to destructive incorporation into the human sphere, they may join a streamwatch group or a group to protect local wildlife, for example.

Instrumentalism (the fifth feature) involves the assumption that all other species are available for unrestricted human use, although it is unlikely that many of those steeped in the ideology of human supremacy will see humans as similarly available for non-human use (for example, as food). Instrumentalism in this form is a clear expression of anthropocentrism and of an arrogant attitude to the other which sees it in the guise of a servant of the self. One of the most important things to aim to establish in any strategy for countering instrumentalism, then, is some degree of human humility and sensitivity to nature's own creativity and agency. Another very important strategy here is the cultivation or recovery of ways of seeing beings in nature in mind-inhabited ways as other centers of needs and striving, to replace the dominant view of them as mere mechanical resources for the use of the center that is our own self. All the above forms of activism may be mobilized in these tasks. But it is often also important to demonstrate the imprudence of anthropocentrism, for example by showing the extent of uncertainty and the limits of our knowledge. This strategy may be especially important where anthropocentrism takes the form, as it often does, of arrogance wrapped in the garments of science. Narrowly anthropocentric cognitive and aesthetic relations to nature can be countered in a variety of ways: for example, promoting alternative caring and attentiveness toward the land, learning about nonanthropocentric models other cultures may be using, and generating local earth narratives which can place local relationships with nature in a deeper, more storied and less narrowly productivist framework of attachment.

This list of ways to counter anthropocentrism in the sense I have outlined is, of course, very much what ecological education and nitty-gritty grassroots environmental activism is all about. To someone looking for cosmic transformation, this account of counteranthropocentric activism may seem disappointingly ordinary. In fact, however, it is precisely this ordinariness that gives it its practical power. It is a model of anthropocentrism that can draw together, deepen and help explain the basis for activist practice, providing a useful foundation for the critical and self-reflective practice we need. It turns out that ecological activists are already doing what is necessary to counter the historical legacy of human-centeredness. Not only is it not impossible, it is going on right now all around us. If this kind of activist practice "walks the talk" of the critique of human-centeredness, the claims of the critics of environmental philosophy that its critical engagement with the philosophical framework of anthropocentrism is misguided, impractical and dispensable for environmental struggle are clearly wrong.[36]

8. The Blindspots of Anthropocentrism

Another vital feature of the Othering model of human-centeredness is that it validates the ecological insight that a human-centered framework is a serious *problem*, and does so in a far clearer way than the cosmic model. The centric structure imposes a form of rationality, a framework for beliefs, which naturalizes and justifies a certain sort of self-centeredness, self-imposition and dispossession, which is what eurocentric and ethnocentric colonization frameworks as well as androcentric frameworks involve. The centric structure accomplishes this by promoting insensitivity to the Other's needs, agency and prior claims as well as a belief in the colonizer's apartness, superiority and right to conquer or master the Other. This promotion of insensitivity is in a sense its function. Thus it provides a wildly distorted framework for perception of the Other, and the project of mastery it gives rise to involves dangerous forms of denial, perception, and belief which can put the centric perceiver *out of touch* with reality about the Other. The framework of centrism does not provide a basis for sensitive, sympathetic or reliable understanding and observation of either the Other or of the self. Centrism is (it would be nice to say "was") a framework of moral and cultural blindness.

Think, for example of what a eurocentric framework led colonizers such as Australians in the past to believe about indigenous people: that they were semi-animals, without worthwhile knowledge, agriculture, culture, or technology, that they were wandering nomads with no ties to the land, and were without religion.

Colonizers believed, despite the existence of over three hundred indigenous languages, in a simple, uniform indigenous character. They failed completely to understand the relationship between Aboriginal people and the land they took, or to recognize indigenous management practices. The eurocentric framework told white Australians that the Aboriginal presence imposed no limits on their actions, that the land was *terra nullius*, simply "available for settlement." Thus, it created a belief system which was the very opposite of the truth, and evidence to the contrary was simply not observed, was discounted or denied. As a number of feminist thinkers have noted in the case of scientific observation, a framework of perception and reason designed for subjugating and denying the other is not a good framework for attentive observation and careful understanding of that other, and even less is it one for evolving life strategies of mutual benefit or mutual need satisfaction.

If human-centeredness similarly structures our beliefs and perceptions about nature, it is a framework for generating ecological denial and ecological blindness in just the same way. The upshot of such a structure in the case of nature is insensitivity to the intricate patterns and workings of nature, encouraging those who hold it to see only a disorderly sphere in need of the imposition of rational order via human development. In this framework, the Other has legitimacy only as a form or servant of the self, and cannot be truly encountered. Since mind is denied to everything but the human, a human-centered framework generates a mechanistic conception of the world which is unable to see in nature other centers of striving and needs for earth resources, just as it is unable to see its own embeddedness in what it destroys.

The human-centered framework is insensitive to the Other's needs and ignores the limits they impose, pursuing an aggressive self-maximization. Just how aggressive this is and how little space it leaves for the other can be seen from the way animals are treated in the name of rational agriculture, with chickens and calves held in conditions so cramped that in a comparable human case they would clearly count as torture. Its logic of the One and the Other tends through incorporation to represent the Other of nature en-

tirely in terms of human needs, as involving replaceable and interchangeable units answering to these needs, and hence to treat
nature as an infinitely manipulable and inexhaustible resource.

But the feature that makes this human-centered framework of
rationality most dangerous of all is that it encourages a massive *denial of dependency*, fostering the illusion of nature as inessential and
leaving out of account its irreplaceability, non-exchangeability and
limits. As part of its historic denial, the human-centered framework
backgrounds and fails to understand the complexity or importance
of the biospheric services provided by global ecosystem processes,
at the same time as it overestimates its own knowledge and capacity for control in a situation of limits. We can see the signs of this
overestimation in various recent events, ranging from the collapse
of fisheries across the world to the failure of Biosphere 2.[37] This denial of dependency combines with the idea of human hyper-separation to promote the illusion of the authentically human as outside
nature, invulnerable to its woes. A framework which is unable to
recognize in biospheric nature a unique, non-tradeable, and irre-
placeable sustaining other on which all life on the planet depends
is deeply anti-ecological. That is why the development of ecological worldviews has been so profoundly revolutionary.

Put this distorting framework of ecological denial beside the reality of our total dependency on the biosphere and the reality of
the present human level of resource use, in which human activity
consumes as much as 40 percent of the net photosynthetic product of the earth, in a pattern which has been doubling every
twenty-five to thirty years.[38] Our species seems to be aiming to di-
vert most of this energy for its own purposes, increasingly requi-
sitioning the resources that biospheric others need to survive. And
thus those processes that maintain the planetary biospheric systems we take for granted are being rapidly and indiscriminately
overridden by the very different structures of human society.

This human-centered framework may once have been functional
for the dominance and expansion of Western civilization, since it
removed the constraints of respect for nature that might otherwise
have held back its phenomenal conquests. But in the age of ecological limits we have now reached, it is highly dysfunctional, and

the insensitivity to the entire biosphere that it promotes is a grave threat to our own as well as other species' survival. The old anthropocentric model that binds our relationships with nature within the logic of the One and the Other prevents us from moving on to the new mutualistic and communicative models we now so urgently need to develop for both our own and nature's survival in a new age.

9. Conclusion: Prudence, Survival, and the Return of Ann and Bruce

Suppose that instead of leaving right away, Ann persuades Bruce to try a visit to a marriage counselor to see if Bruce can change enough to save their relationship. After listening to their stories, the counselor diagnoses Bruce as a textbook case of egocentrism, an individual version of the centeredness structure set out above. Bruce seems to view his interests as somehow radically separate from Ann's, so that he is prepared to act on her request for more consideration only if she can show he will get more pleasure if he does so, that is, for instrumental reasons which appeal to a self-contained conception of his interests. He seems to see Ann not as an independent person but as someone defined in terms of his own needs, and claims it is her problem if she is dissatisfied or miserable. Bruce sees Ann as there to service his needs, lacks sensitivity to her needs and does not respect her independence or agency.[39] Bruce, let us suppose, also devalues the importance of the relationship, denies his real dependency on Ann, backgrounds her services and contribution to his life, and seems to be completely unaware of the extent to which he might suffer when the relationship he is abusing breaks down. Bruce, despite Ann's warnings, does not imagine that it will, and is sure that it will all blow over: after a few tears and tantrums Ann will come to her senses, as she has always done before, according to Bruce.

Now the counselor, Jane, takes on the task of pointing out to Bruce that his continued self-centeredness and instrumental treatment of Ann is likely to lead in short order to the breakdown and loss of the relationship. The counselor tries to show Bruce that he has underestimated both Ann's determination to leave unless there is change, and the sustaining character of the relationship. Jane points out that he may, like many similar people the counselor has seen, suffer much more severe emotional stress than he realizes when Ann leaves, as she surely will unless Bruce changes. Notice, then, that Jane's initial appeal to Bruce is a prudential one; Jane tries to point out to Bruce that he has misconceived the relationship and to make him understand where his real interests lie. There is no inconsistency here; the counselor can point out these damaging consequences of instrumental relationship for Bruce without in any way endorsing or encouraging purely instrumental relationships.

In the same way, the critic of human-centeredness can say with perfect consistency, to a society trapped in the centric logic of the One and the Other in relation to nature, that unless it is willing to give enough consideration to nature's needs, it too could lose a relationship whose importance it has failed to understand, has systematically devalued and denied—with rather more serious consequences for survival than in Bruce's case. The account of human-centeredness I have given, then, unlike the cosmic account demanding self-transcendence and self-detachment, does not prohibit the use of certain forms of prudential ecological argument, although it does suggest certain contexts and limits to their use.

In the case of Ann and Bruce, Jane might particularly advance these prudential reasons as the main reasons for treating Ann with more care and respect at the initial stages of the task of convincing Bruce of the need for change. And prudential arguments need not just concern the danger of losing the relationship. Jane may also try to show Bruce how the structure of egocentrism distorts and limits his character and cuts him off from the main benefits of a caring relationship, such as a sense of the limitations of the self and its perspectives obtained by an intimate encounter with someone else's needs and reality. Prudential arguments for respect are the kinds of arguments that are especially useful in an *initial* context of denial, while there is still no realization that there is a serious problem, and resistance to undertaking work for change. In

the same way, the appeal to prudential considerations about ecological damage to humans is especially appropriate in the initial
context of ecological denial, where there is still no systematic ac-
knowledgement of human attitudes as a problem, and resistance to
the idea of undertaking substantial social change.

But once Jane's prudential argument has broken Bruce's initial
resistance to considering change, Jane can and should go on to supplement these prudential arguments for considering Ann, framed
in terms of disadvantages for Bruce from failing to do so, with further kinds of considerations which treat outcomes for Ann as moral
reasons in their own right. Only when he does so can Bruce fully
encounter Ann as another person, an equal moral center, and only
then will he really begin to realize the full benefits of the relationship, for the full rewards of personal relationships of care only
come when we have ceased to be primarily focussed on the benefits we ourselves may gain from them, and are absorbed in the
other. These further considerations Jane introduces then are not
primarily oriented to outcomes for Bruce: that is, they are not instrumental. It is here, then, that Jane can open up the couple's exploration of the strategies for respecting, negotiating and balancing the needs of self and other. These will probably have to
emphasize improved communication, about mutual emotions,
needs, desires and limits. They may involve, especially for Bruce,
a more mutualistic reconception of self and self's needs as a self-
in-relationship, formed in a balance of mutual transformation
rather than in a context where Bruce is always the controller or
transformer and Ann the one transformed. But although the ad-
vantages or disadvantages to Bruce cannot be the only kinds of
considerations Jane introduces to Bruce if she is aiming to help
Ann re-negotiate the relationship in a framework that is respect-
ful of Ann, Jane does not have to exclude these questions or Bruce's
needs from consideration either, at any stage from this process of
re-negotiation, as the fallacious view of prudence as always instrumental and egocentric suggests.

For example, it would be prudent for Ann to want to be assured
that consideration of her needs will be a settled feature of Bruce's
behavior towards her, to ask for some real assurance that he will

continue considering her needs. So Ann will probably want to be sure Bruce is now acting out of deeper underlying reasons of care, truly considering her needs apart from his own, rather than the sorts of narrowly instrumental reasons Bruce puts up to her in their earlier dialogues.[40] Although there is a basic opposition between the exclusively instrumental mode which reduces the other to a means to the self's ends, and the respect/care mode which acknowledges the other as a different center of agency and value, prudential reasons can quite properly supplement and balance a care perspective, and care itself must have prudential aspects.[41]

The case of Ann and Bruce is meant to illustrate the role of prudence in maintaining durable ethical relationships. But of course the sorts of prudential reasons for considering nature I outlined earlier which invoke the threat or danger from the ecological blindspots induced by anthropocentrism are by no means the only kinds of reasons the Othering model suggests for regarding anthropocentrism as a prudential liability. Another very important set of reasons why human-centeredness is a problem derives not directly from its ill effects on the colonized, on animals and nature themselves, but more indirectly from its distorting effect on the colonizers, on human identity, and human society. The structure of human-centeredness distorts and limits the possibilities for *what we can become as humans* in much the same way that the structures of racism and sexism do for colonizer identities and for masculinist identities, and the structure of egocentrism does for Bruce. The logic of human-centeredness which conceives nature, external and internal, as the "lower," denied aspect of self and society, also constructs dominant human identity and virtue as the identity and virtue of the master, by the exclusion of othered aspects of identity both within and without.[42]

This kind of structure prevents the formation of properly integrated selves and of certain beneficial and satisfying kinds of relationships with others and with nature. Again, just as we do not realize the benefits of personal relationships of care until we have ceased to be primarily motivated by or focussed on the benefits we gain from them, so we can realize fully the rewards of experiencing the other of nature as another center (or rather, many centers) only when our primary focus is not our own gain or even safety. And to the extent that anthropocentric frameworks prevent us from experiencing the others of nature in their fullness, we not only imperil ourselves through loss of sensitivity but also deprive ourselves of the unique kinds of richness and joy the encounter with the more-than-human presences of nature can offer. To realize this potential, we will need to reconceive the human self in more mutualistic terms, as a self-in-relationship with nature, formed not in the drive for mastery and control of the other but in a balance of mutual transformation and negotiation. In this way we can begin to replace the old instrumental and mechanistic models which guided the development of the West during the centuries of conquest, and to meet the challenge of our time to realize the new models of communication and care which are now struggling to emerge.

Notes

1. John Passmore, *Man's Responsibility for Nature* (London: Duckworth, 1974, 2nd ed. 1980). Don Mannison, "What's wrong with the concrete jungle?," In: D. Mannison, et al (eds.), *Environmental Philosophy* (Canberra: ANU, 1980).

2. Andrew Dobson, *Green Political Thought*. (London: Routledge, 1990). Brian Norton, "Environmental Ethics and Weak Anthropocentrism," *Environmental Ethics* (6) (1983), 211–224. Brian Norton, *Towards Unity Among Environmentalists* (New York: Oxford University Press, 1991).

3. William Grey, "Anthropocentrism and Deep Ecology," *Australasian Journal of Philosophy* 71 (1993), pp. 463–475. For an earlier argument see Janna Thompson, "A Refutation of Environmental Ethics," *Environmental Ethics* 12 (1990), pp. 147–160.

4. Ibid., p. 473.

5. Thus, Grey writes "we [must] eschew all human values, interests and preferences." Ibid., p. 473.

6. According to Kant ("Duties to Animals and Spirits," in *Lectures on Ethics*, trans. Louis Infield (New York: Harper and Row, 1963), pp. 239–241), for example, non-humans could be taken into consideration only *indirectly*, to the extent that damaging them also damages humans. The beating of a dog, according to Kant, was not wrong because of injustice to the dog, but because it developed cruelty in humans which would then be exercised *directly* on humans. One catch with this approach is that in the case where we are able to find some way to stop the cruelty to the dog spilling over into cruelty to humans, or if cruelty to the dog substituted for cruelty to humans, Kant would be able to find nothing at all wrong with cruelty to the dog. If it substituted for cruelty to humans, I suppose, it could actually be a duty!

7. Iris Young, *Justice and the Politics of Difference.* (Princeton: Princeton University Press, 1991), p. 105.

8. Joseph Butler, "Upon the Love of Our Neighbour," in *Fifteen Sermons upon Human Nature* (1726, London 2d ed. 1729). Reprinted in Ronald D. Milo (ed.), *Egoism and Altruism* (Belmont CA: Wadsworth Publishing Co., 1973), pp. 26–36.

9. A polycentric world has many centers, whereas an acentered (decentered) one has none. Depending on how it is developed, a sufficiently inclusive and flexible polycentrism might also be described as a (relatively) de-centered world. See Molefi Kete Asante, *The Afrocentric Idea* (Philadelphia: Temple University Press, 1987). Edward Said, *Orientalism* (New York: Vintage, 1979).

10. Molefi Kete Asante, *The Afrocentric Idea* (Philadelphia: Temple University Press, 1987).

11. Ama Mazama, "The Relevance of Ngugi Wa Thiong'o for the African Quest," *The Western Journal of Black Studies,* Vol. 18, No. 4 (1994), pp. 211–218.

12. Catherine Mackinnon, *Feminism Unmodified* (Cambridge: Harvard, 1987). Iris Young, *Justice and the Politics of Difference* (Princeton: Princeton University Press, 1991).

13. Nancy Hartsock, "Foucault on Power: A Theory for Women?," in L. Nicholson, (ed.), *Feminism/Postmodernism* (New York: Routledge, 1990).

14. Michele le Doeuff, *Hipparchia's Choice* (London: Routledge, 1989).

15. See especially Simone de Beauvoir, *The Second Sex* (London/New York: Foursquare Books, 1965). Nancy Hartsock, "Foucault on Power: A Theory for Women?," in L. Nicholson (ed.), *Feminism/Postmodernism*. (New York: Routledge, 1990). Marilyn Frye, *The Politics of Reality* (New York: The Crossing Press, 1983).

16. I use the capitalized terms "One" and "Other" to denote forms of otherness which exhibit the logic of the One and the Other I outline here.

17. Any institutionalized system of domination that aims to avoid arbitrary elements and take full advantage of cultural potential for its reproduction must aim to separate the dominating group from the others, and will tend to adopt cultural means which define the identity of the center (usually cast in the West as Reason) by exclusion of the inferiorized qualities of the Other. Hyper-separation maximises security for the dominating group.

18. On such a concept of "race," see Stephen Jay Gould, *The Mismeasure of Man* (New York: W.W. Norton, 1981).

19. On this point see especially Gloria Marshall, "Racial Classifications: Popular and Scientific," in Sandra Harding (ed.), *The Racial Economy of Science* (Indianapolis: Indiana University Press, 1993), pp. 116–27; and Rhett C. Jones, "The End of Africanity? The Bi-racial Assault on Blackness," *The Western Journal of Black Studies*, Vol. 18, No. 4 (1994), pp. 201–10. Exclusionary motives often generate absurdly fine distinctions in order to maximize separation and maintain the image of discontinuity, such as the "half-caste" and "quarter-caste" distinctions of Australia and the United States "one drop" rule.

20. Nancy Hartsock, "Foucault on Power: A Theory for Women?," in L. Nicholson, (ed.), *Feminism/Postmodernism* (New York: Routledge, 1990).

21. Nancy Leys Stepan, "Race and Gender: the Role of Analogy in Science," in Sandra Harding (ed.), *The Racial Economy of Science* (Indianapolis: Indiana University Press, 1993), pp. 359–76.

22. See Albert Memmi, *The Coloniser and the Colonised* (New York: Orion Press, 1965).

23. Edward Said, *Orientalism* (New York: Vintage, 1979).

24. Marilyn Frye, *The Politics of Reality* (New York: The Crossing Press, 1983), p. 32.

25. See Ama Mazama, "The Relevance of Ngugi Wa Thiong'o for the African Quest," *The Western Journal of Black Studies*, Vol. 18, No. 4 (1994), pp. 211–18.

26. Simone de Beauvoir, *The Second Sex* (London/New York: Foursquare Books, 1965), p. 8.

27. Benita Parry, "Problems in Current Theories of Colonial Discourse," in Bill Ashcroft, Gareth Griffiths, and Helen Tiffin (eds.), *The Post-Colonial Studies Reader* (Routledge: London, 1995), p. 42.

28. See Albert Memmi, *The Coloniser and the Colonised* (New York: Orion Press, 1965).

29. On disvaluation, see Anthony Weston, "Self-Validating Reduction: Toward a Theory of Environmental Devaluation," *Environmental Ethics* 18 (1996), pp. 115–32.

30. On indigenous resistance in the Australian case see Henry Reynolds, *The Other Side of the Frontier: Aboriginal Resistance to the European Invasion of Australia* (Melbourne: Penguin, 1982); and for the American case see Frederick Turner, *Beyond Geography: The Western Spirit Against the Wilderness* (New Brunswick, NJ: Rutgers University Press, 1986).

31. A number of ecofeminists have remarked on this feature. See for example Rosemary Radford Ruether, *New Woman New Earth* (Minneapolis, MN: Seabury Press, 1975); and Karen Warren, "The Power and Promise of Ecological Feminism," *Environmental Ethics* 12 (1990), pp. 121–46.

32. See Nancy Leys Stepan, "Race and Gender: the Role of Analogy in Science" in Sandra Harding (ed.), *The Racial Economy of Science* (Indianapolis: Indiana University Press, 1993), pp. 359–76.

33. Stephen Jay Gould, *The Mismeasure of Man* (New York: W.W. Norton & Co., 1981).

34. Another closely related source of transfer of these metaphors and models of identity is the rationalist development of the concept of reason in relation to its contrast class of nature. See Val Plumwood, *Feminism and the Mastery of Nature* (London: Routledge, 1993).

35. In the new paradigm of modernity, control of nature will provide freedom from necessity in the form of the body and death in this world, rather than in a transcendent afterlife as in the paradigm of Christianity it supplanted.

36. I am not suggesting that every form of environmental activism engages fully with anthropocentrism. There can of course, as in all other areas of social change, be more and less thorough forms of environmental understanding and challenge, deeper forms which look at longer-term solutions and consider problems in more unified and systematic ways, and shallower forms content with quick-fixes which consider only a restricted range of affected interests and problems.

37. See David Tilman and Joel E. Cohen, "Biosphere 2 and Biodiversity," *Science* 274 (1996), pp. 1150–51.

38. See Susan George, *The Debt Boomerang* (London: Pluto, 1993).

39. For many psychological theorists, this is also an account of the masculinised self. See for example Marilyn Frye, *The Politics of Reality* (New York: Crossing Press, 1983); and Nancy Chodorow, "Gender, Relation and Difference in Psychoanalytic Perspective," in H. Eisenstein and A. Jardine (eds.), *The Future of Difference* (New Brunswick, NJ: Rutgers, 1985), pp. 3–19.

40. So-called enlightened self-interest, often appealed to as the savior here, is only as good as the assurance that the actor will remain enlightened, and there is a major question of what can guarantee that regularly in the absence of a dispositional base such as care.

41. On the importance of prudence in a feminist context, see Jean Hampton, "Selflessness and Loss of Self," in E.F. Paul, F.D. Miller, and J. Paul (eds.), *Altruism* (Cambridge: Cambridge University Press, 1993), pp. 135–65.

42. See Val Plumwood, *Feminism and the Mastery of Nature* (London: Routledge, 1993).

Ethics on the Home Planet

Holmes Rolston, III

Views of Earth from space are the most impressive photographs ever taken, if one judges by their worldwide impact. They are the most widely distributed photographs ever, having been seen by well over half the persons on Earth. Few are not moved to a moment of truth, at least in their pensive moods. The whole Earth is aesthetically stimulating, philosophically challenging, and ethically disturbing. The world view is an invitation to environmental philosophy. "Once a photograph of the Earth, taken from *the outside* is available . . . a new idea as powerful as any in history will be let loose."[1] The call is to rethink an emerging vision of Earth and the place of human life upon it. The distance lends enchantment, brings us home again. The distance helps us to get real. We humans get put in our place. We ask what we ought to do.

A virtually unanimous experience of the nearly two hundred astronauts, from many countries and cultures, is the awe experienced at the first sight of the whole Earth—its beauty, fertility, smallness in the abyss of space, light and warmth under the sun in surrounding darkness and, above all, its vulnerability. In the words of Edgar Mitchell, Earth is "a sparkling blue-and-white jewel . . . laced with slowly swirling veils of white . . . like a small pearl in a thick sea of black mystery."[2] "I remember so vividly," said Michael Collins, "what I saw when I looked back at my fragile home—a glistening, inviting beacon, delicate blue and white, a tiny outpost suspended in the black infinity. Earth is to be treasured and nurtured, something precious that *must* endure."[3] There is a vision of an Earth ethic in what he sees.

The two great marvels of our planet are life and mind, both among the rarest things in the universe; so far really unknown elsewhere. In the global picture, the late-coming, moral species, *Homo sapiens*, arising a few hundred thousand years ago, has, still more lately in this century, gained startling powers for the rebuilding and modification, including the degradation, of this home planet. The four most critical issues that humans currently face are peace, population, development, and environment. All are entwined. Human desires for maximum development drive population increases, escalate exploitation of the environment, and fuel the forces of war. Those who are not at peace with one another find it difficult to be at peace with nature, and vice versa. Those who exploit persons will typically exploit nature as readily—animals, plants, species, ecosystems, and Earth itself.

So we are searching for an ethics adequate to respect life on this Earth, the only planet yet known with an ecology. On Earth, home to several million species, humans are the only species with moral responsibilities of this kind. Earth is the only planet "right for life," and ethics asks about the "right to life" on such a planet. Certainly it seems "right" that life continue here; life is, in the deepest sense, the most valuable phenomenon of all, with its prolific history since its origin three and a half billion years ago.

Socrates said, famously, as an invitation to philosophy, "The unexamined life is not worth living." To that I wish to add, in invitation to environmental philosophy: "Life in an unexamined world is not worthy living either." We miss too much of value. For the trip you are about to take I offer myself, you might say, as a wilderness guide. A century ago the challenge was to know where you were geographically in a blank spot on the map, but today we are bewildered philosophically in what has long been mapped as a moral blank space. Values run off our maps. We are beginning to see that we cannot figure out who we are until we know where we are, a unique species, *Homo sapiens*, the wise species, on a unique Earth. Philosophy is the love of wisdom (*philos*, loving; *sophia*, wisdom); caring for the Earth has become vital in that quest.

1. Ethics for People

Well, yes, you may be saying, Earth is an impressive planet, but ethics is for people. People are both the subject and the object of ethics, in the sense that only humans are deliberative moral agents and also that humans have obligations only to other humans. Humans are helped or hurt by the condition of their environment, and that is what environmental philosophy is about.

Or so it might first appear, a truth which (we will argue) is only a half truth. Humans can and ought to be held responsible for what they are doing to their Earth, which is their life support system. That is true enough. We are not responsible, of course, for Earth's being here past and present; we are latecomers in evolutionary history. But we are becoming increasingly responsible for Earth's future. In that sense, everything humans value is at stake in seeking sustainable development, a sustainable biosphere. If there are any duties at all, we must care for this surrounding world, since this is the home for us all. But, so this argument goes, these are duties owed by people to other people (as well as to themselves); caring for the planet is a means to this end.

One sometimes encounters claims that chimpanzees or other animals have the precursors of moral behavior, and that may be so. But it is pointless to invite the bears or daisies to do environmental philosophy; they are incapable of it. One is mistaken to blame wolves for killing sheep or to censure weeds for growing in the wrong place. They are doing what comes naturally, instinctively, about which they have few, if any, options and choices. They really cannot do otherwise, and so cannot be held to account morally. Maybe there are some animal virtues, but they are not in doing philosophy and ethics. The natural world is amoral; morality appears when humans in their cultures emerge out of previous nature. Our biochemistries are natural, to be sure, and humans draw their life support from the hydrological cycles and photosynthesis; humans too have genes and inborn traits; they are subject to natural laws.

But human life is radically different from that in spontaneous nature. Unlike coyotes or bats, humans are not just what they are by nature; we come into the world by nature quite unfinished and become what we become by culture. Animals are often social, of course; they can imitate the behaviors of parents or others in their packs or flocks, as when birds learn the migration routes by following others. Animal behavior is not always genetically stereotyped; it may be labile, subject to development only if environmental circumstances require or permit it, including their social interactions. But none of these hereditary factors resembles a cumulative transmissible culture, as is so strikingly present in humans. The determinants of animal and plant behavior are never anthropological, political, economic, technological, scientific, philosophical, ethical, or religious.

With their culturally formed worldviews, humans deliberately and extensively rebuild the spontaneous natural environment and make the rural and urban environments in which they reside. It is the quality of life in these environments, hybrids of nature and culture, about which we should care—so this people-centered argument runs. Ethics arises to protect various goods within our cultures; this, historically, has been its principal arena. As philosophers frequently model this, ethics is a feature of the human social contract.

If ethics is in any sense natural to humans, it will be in some special nature of this highly intelligent, quite social species, in *human* nature more than in *wild* nature. If ethics is rational for humans, this will be because there are benefits to persons who live in the resulting kind of culturally constructed society. Natural processes in the wild serve animal life well; but these are not processes in term of which the values achieved in culture can be fully protected; and, one way or another, there emerges morally responsible agency to protect human life and its cultural values.

People arrange a society where they and the others with whom they live do not (or ought not) lie, steal, kill, and so on. This is right, and one reason it is right is that people must cooperate to survive; and the more they reliably cooperate the more they flourish. In this social contract, each must respect the goods of the others. There is no reason for others to tell me the truth, respect my life and property, and so on, unless I reciprocate. This will be true for everybody. So it is in everybody's best interest to enter into this social contract.

There will be tradeoffs, my good against yours, and hence the sense of justice arises (each his or her due), or fairness (equitable outcomes for each), or of greatest good for greatest number. Such standards can appeal to every actor, in whatever culture (though the detailed content will to some extent be culturally specific), because on the whole this is the best bargain that can be struck, mindful of the required reciprocation. Human well-being depends on it. Further, there is considerable satisfaction both in being fairly treated and in realizing that you keep your end of the bargain, even at some cost. Further still, one's identity and interests get vested in other people and causes, with which one shares one's values. What one ought to do, in any place, at any time, whoever one is, is what optimizes humanly shared values, and this is generically good, both for the self and all others.

Beginning with a sense of one's own values to be defended, ethics requires becoming more inclusive, recognizing that one's own self-values are widely paralleled, a kind of value that is distributed in myriads of other selves. The defense of one's own values gets mixed, willy-nilly, with the defense of the values of many others. One way of envisioning this is the so-called "original position," where one enters into contract, figuring out what is best for a person on average, oblivious to the specific circumstances of one's time and place. This is where the sense of universality, or at least panculturalism, in morality has a plausible rational basis.

But the problem with animals, much less plants, or species or ecosystems, or mountains and rivers, is that they are out of all this. They are not such contracting parties at all. They cannot be held responsible, nor does their flourishing depend on any such reciprocating in our human society. We cannot invite them to do environmental philosophy, or be ethical. So it may seem that ethics stops with humans in their cultures. Man is the measure of things, said Protagoras, an ancient Greek philosopher, setting the tone of philosophy since. Humans are the measures, the valuers of things, even when we measure what they are in themselves.

John Passmore thinks that only paradigmatic human communities generate obligations:

> Ecologically, no doubt, men do form a community with plants, animals, soil, in the sense that a particular life-cycle will involve all four of them. But if it is essential to a community that the members of it have common interests and recognize mutual obligations, then men, plants, animals, and soil do *not* form a community. Bacteria and men do not recognize mutual obligations, nor do they have common interests. In the only sense in which belonging to a community generates ethical obligations, they do not belong to the same community.[4]

Passmore is assuming that the members of a morally bound community must recognize reciprocal obligations. If the only communal belonging that generates obligations is this particular social sense, involving mutual recognition of interests, then the human community is the sole matrix of morality, and environmental ethics is a nonstarter. So unless we can find a revised concept of biotic community and of what duties can be toward, of what values are to be measured, there will be no duties to the environment.

Although we go on (in the next sections) and ask about duties to others than humans, a great deal of the work of environmental ethics can certainly be done from within the social contract. Humans need to be healthy, for instance; physical health is as much part of their biology as mental health is part of their culture. Health, however, is not simply a matter of biology from the skin–in. Environmental health, from the skin–out, is just as important. In what they do concerning the natural world, some actions are healthy for humans because they agree with the ecological processes with which their cultural decisions interact. After all, humans too, like the animals and plants, need reasonably clean air and water. Even in agriculture, humans must grow their food in some soil that is more or less unpolluted (use pesticides and herbicides though they may) and fertile (use fertilizers though they may). It is hard to have a healthy culture in a sick environment.[5]

Nor is environmental health just minimal; think rather of a quality environment. Humans enjoy the amenities of nature— wildlife and wildflowers, scenic views, places of solitude—as well as the commodities—timber, water, soil, natural resources. This is part of our mental health. Supporting environmental health and a quality environment can certainly be counted as duties within a social contract. So, sometimes at least, decisions in environmental ethics and in social ethics can be win-win. There are nonrival, complementary goods. Properly to care for the natural world can combine with a strategy for sustainability. If nature provides the life support system for culture, what is good for nature is often good for culture.

Environmental ethics, by this account, is founded on what we might call a human right to nature.[6] Such a right includes the basic natural givens—air, soil, water, functioning ecosystems, hydrologic cycles, and so on. This right has not figured much in the heritage of our past, because it could previously be taken for granted. But now it must be made explicit, and defended. If humans have a right to life, liberty, and the pursuit of happiness, then they have a right to the natural conditions that are instrumental to produce this.

Once we had to defend the right to own property, to vote, to a basic education. We have discovered of late there is one more domain where humans have fundamental values at stake, always present but only recently consciously appreciated, a domain so threatened that it must come under new social protection. The "right to a quality environment" has been proposed as constitutional amendment in the United States by the National Wildlife Federation. Some nations that have recently rewritten their constitutions have stipulated such a right. The World Commission on Environment and Development has proposed: "All human beings have the fundamental right to an environment adequate for their health and well-being."[7]

Such a "right to nature" is a right within culture, that is, it is a claim we can make against intrusions by other humans where these put a healthy environment in jeopardy. We all have a right to object when other humans foul the air, destroy the soils, drive other lifeforms to extinction, burn the rainforest. Asserting this right this does not mean that humans have some kind of claim against Mother Nature itself, for nature is no moral agent. We cannot lay claims against grizzly bears or wildflowers, rivers or mountains. There is no right to be claimed against nature for these processes and products. Nature is prolific but not responsible. In fact, if we leave the human world, the social contract, and turn to nature as it is independently of the human presence, things change—dramatically.

Ethics is for people, but is ethics only about people? What has ethics to say about the rest of life on our planet? Environmental philosophy is rather new, at least in its current form, although people have thought about nature across many centuries. This can seem rather strange, since we often think that today the scientific accounts of nature are better than ever. More than any other people who have previously lived on the planet, thanks to modern physics, chemistry, geology, meteorology, biology, including evolutionary and ecological science, as well as biochemistry, we have accurate descriptions of nature. True, we have a lot yet to learn, and natural systems have proved to be more open and complex than we thought. Still, we know an enormous amount. But the problem is that, despite all these scientific accounts, nature has been mapped philosophically as a moral blank space, as value-free in and of itself.

Trying to map the human environments, we are valuing three main territories: the urban, the rural, and the wild—all three of which are necessary if we are to be three-dimensional persons. In some parts of the world (such as Denmark), the environments are either urban or rural; but in many parts of the world (such as Finland), considerable wildness remains. Over the Earth as a whole, in one survey, using three categories, researchers find the proportions of Earth's terrestrial surface altered as follows: (1) Little disturbed by humans, 51.9 percent; (2) Partially disturbed, 24.2 percent; and (3) Human dominated, 23.9 percent. Factoring out the ice, rock, and barren land, which supports little human or other life, the percentages become: (1) Little disturbed, 27.0 percent; (2) Partially disturbed 36.7 percent; and (3) Human dominated 36.3 percent.[8] Most terrestrial nature is dominated or partially disturbed (73.0 percent). Still, nature that is little or only partially disturbed remains 63.7 percent of the habitable Earth. Also, of course, there is the sea, less affected than the land; and the oceans cover most of the Earth.

Turning to such nature, one approach is to see ourselves, humans, as valuing our roots, and our neighbors and others, who live along with us in the partially disturbed environments, who are present and flourish even more in the little disturbed places. The first two terms indicate kinships, but the third term goes further, to discover genuine "others," even "aliens," to put the point with some force. If ethics is not just for people, we humans must cross over diffuse boundaries into regions shared by neighbors and later we travel still further from our humankind, from our home territory. *Homo sapiens*, a unique species, is only one among five million (or ten or twenty, we do not yet know how many) species on Earth, only one among five billion (or ten, or more) species that have come and gone over evolutionary history.

The challenge here for environmental philosophy is how to get people, who perhaps alone on the planet can be ethical, to care for a world that is our home planet and yet also the home for these other creatures. Nature is, of course, much present in the hybrid habitats of rural landscapes, and even in our cities. Here we will also need an ethic for domesticated animals, such as livestock and pets, animals for whom humans have undertaken a responsibility and may be putting to some resourceful use. Those issues are the concern of animal welfare ethics. Also, wildlife can still extensively remain on landscapes put to multiple use; and so we need an ethic of wildlife management.

Wild nature—unmanaged nature in the spontaneously wild sense—is part of our global environment, and yet not our human habitat. Examples are wilderness areas and nature reserves, which we try to manage as little as possible, or to manage human uses of them as far as we can to let nature take its course. The wild is an environment that humans need and ought to respect, but it is not an environment in which we can reside and still be human. "Man is by nature a political animal," said Aristotle—the animal who builds and inhabits a "polis," a town. *Homo sapiens* is the one species that rebuilds its environment on the basis of a cumulative transmissible culture. Man is generically an animal, but specifically a citizen; that specific characteristic identifies the human essence. That is why, some say, ethics arises to govern conduct in the "polis," with its social contract, channeling, orienting behavior to protect the goods of human nature and culture. Now, however, we are beginning to recognize that the larger realm of nature encompasses the city too. Likewise our ethical view must now grow more encompassing as well—without leaving the city behind.

2. Ethics for Animals

Domesticated animals, as we were saying, are hybrids, almost artifacts, since they have been bred so carefully. Their lives are under human control; they hardly have lives of their own, certainly not on their own, and their mixed status is problematic. They need protection, owing to their compromised nature, but we cannot formulate an adequate environmental ethic on the basis of our obligations to livestock, laboratory animals, and pets, since what they are is so largely the measure we make of them. An animal welfare ethic will not be the same as an environmental ethic. Mary Midgley calls these relationships those in "mixed communities."[9] The challenge is to constrain an inevitably anthropocentric community by animal values that are present but admittedly not at the center of the ethical focus.

We can bring animals more directly into focus by considering wild animals. They do not make man the measure of things at all. There is no better evidence of nonhuman values and valuers than spontaneous wild life, born free and on its own. Animals hunt and howl, find shelter, seek out their habitats and mates, care for their young, flee from threats, grow hungry, thirsty, hot, tired, excited, sleepy. They suffer injury and lick their wounds. Here we are quite convinced that value is non-anthropogenic,* to say nothing of anthropocentric. These wild animals defend their own lives because they have a good of their own. There is somebody there behind the fur or feathers. Our gaze is returned by an animal that itself has a concerned outlook. Animals are value-able: able to value things in their world.

Discovering such values forces us to ask whether at least some of what counts in ethics is generic to our kinship with animals, not just specific to our species. First common sense and later science teaches us many similarities of animals and humans; nobody really doubts that animals get hungry and suffer pains for instance. The protein coding sequences of DNA for structural genes in chimpanzees and humans is more than 99 percent identical.[10] Confronted with such facts, we have to philosophize over them.

The first thought seems to be the simple recognition that we are indeed related, kin with others in our biotic community, whether these communities are those of the wild, the rural, or even the suburban. By parity of reasoning, it seems that what we value in ourselves, if we find this elsewhere, we ought also to value over there, in others. There is a sympathetic turning to value what does not stand directly in our lineage but is like enough ourselves that

*Not generated by humans.—Ed.

we are drawn by spillover to shared phenomena manifest in others. The principle of universalizability demands that I recognize corresponding values in fellow persons. Growth in ethical sensitivity has often required enlarging the circle of neighbors to include other races and cultures. But this principle does not apply only with reciprocating moral agents.

Beyond that, animals take an interest in affairs that affect them. A moose does not suffer the winter cold, as we might if there (we having evolved in the tropics); perhaps the warbler is not glad when it sings. But we must not commit the humanistic fallacy of supposing no natural analogues to what humans plainly value. We have every logical and psychological reason to value posit degrees of kinship.

We do not want to be discriminatory, unfair, in ethics, to treat living beings inequitably because we misperceive what values are there. Nor do we want to be undiscriminating, blind to the advanced achievements, to the excellences, even the virtues that are superbly expressed in the animal world. Young and fully of trigger-itch, Aldo Leopold once shot a wolf, mortally wounding her. "We reached the old wolf in time to watch a fierce green fire dying in her eyes. I realized then, and have known ever since, that there was something new to me in those eyes—something known only to her and to the mountain."[11] Two-thirds of a century later, we have put wolves back in the landscape, in Yellowstone National Park, because we want the wisdom we can gain from looking into the fire in those fierce eyes. We have reached the conviction that they, as much as we, belong on the landscape of this Earth we together inhabit.

Nature is often a strange place. Our human roots may lie in wild nature, but wild nature also turns out often to be a radically different place. There are phylogenetic lineages far removed from our own. Here is a new challenge in environmental philosophy. We do not want to measure nonhumans by human standards, though we sometimes want to measure nonhumans and humans by comparable standards. We also frequently run past our capacity to argue by analogy from the value of our experience. For there are quite alien forms of life, with whom we can hardly identify experientially.

Octopus is a mollusc that a primate can recognize as a fellow creature. It is very easy to identify with *Octopus vulgaris*, even with individuals, because they respond in a very "human" way. They watch you. They come to be fed and they will run away with every appearance of fear if you are beastly to them. Individuals develop individual and sometimes irritating traits . . . and it is all too easy to come to treat the animal as a sort of aquatic dog or cat.

Therein lies the danger. It is always dangerous to interpret an animal's reactions in human terms, but with dogs or cats there is a certain reasonableness in doing so. We are mammals too. . . . The octopus is an alien. It is a poikilotherm, never had a dependent childhood, has little or no social life. It may never know what it is to be hungry. . . . The animal, it is true, learns under conditions that would lead to learning in a mammal but the facts that it learns about its visual and tactile environment are sometimes very different from those that a mammal would learn in similar circumstances. Simply because it is evidently intelligent and possessed of eyes that look back at us, we should not fall into the trap of supposing that we can interpret its behavior in terms of concepts derived from birds or mammals. This animal lives in a very different world from our own.[12]

Those who take one evolutionary route in sentient experience are precluded from the direct experience of alien routes, which also have their integrity. Humans can recognize that integrity even though participation in it remains foreign to us. We can grant that the octopus is a center of experience, a subject (while we doubt that a mussel is), and respect a marine lifeform with which we cannot empathize. Some may think it logically or psychologically impossible to value what we cannot share, but this underestimates the human genius for appreciation. Some respect for alien forms of life seems plausible, even if we are slipping away into realms of experience that we cannot reach, and therefore, the critics will say, realms it will be difficult to evaluate.

3. Ethics, Plants, and the Value of Life

A duty to an octopus? Maybe. But can there be duties to a daisy? That claim, many will think, is just too wild. All the familiar moral landmarks are gone. We are not caring about humans, or culture, or moral agents, or animals that are close kin, or can suffer, or experience anything, or are sentient. Plants are not valuers with preferences that can be satisfied or frustrated. But then again, ethics must be about appropriate respect for life, and the higher animals (vertebrates) represent only 4 percent of the living organisms on Earth by species and a minuscule fraction were we to count numbers of individuals. Does the rest of the biosphere count at all in our moral consideration?

A favorite campground in the Rawah Range of the Rocky Mountains is adjacent to subalpine meadows profuse with daisies, lupines, columbines, delphiniums, bluebells, paintbrushes, penstemons, shooting stars, and violets. The trailside signs for years read, "Please leave the flowers for others to enjoy." When I returned once, the wasted wooden signs had been replaced by newly cut ones: "Let the flowers live!" Perhaps the intent was only subtly psychological, but I suspected a shifting ethic; respect for plants replacing what was before only respect for persons.

In the 1880s a tunnel was cut through a giant sequoia in what is now Yosemite National Park. Driving through the Wawona tree, formerly in horse and buggy and later by car, amused millions. The tree was perhaps the most photographed in the world. The giant blew over in the snowstorms of 1968–69, weakened by the tunnel. Some proposed that the Park Service cut another. But the rangers refused, saying that one was enough, and that this is an indignity to a majestic sequoia. It is better to educate visitors about the enormous size and longevity of redwoods, their resistance to fire, diseases, insect pests, better to admire what the stalwart tree is in itself. The comedy of drive-through sequoias perverts the best in persons, who ought to be elevated to a nobler experience. But there is a deeper conviction; using trees for serious human needs can be justified; a silly enjoying of prime sequoias cannot. It perverts the trees.

A plant is a spontaneous life system: self-maintaining with a controlling program (though with no controlling center, no brain). It executes this project, checking against performance in the world, using feedback loops. It composes and recomposes itself, maintaining order against disordering tendencies. Plants do not, of course, have ends-in-view. They are not subjects of a life, and in that familiar sense, they do not have goals. Yet each plant maintains a botanical identity, posting a boundary between itself and its environment. An acorn becomes an oak; the oak stands on its own.

An inert rock exists on its own, making no assertions over the environment. The plant, by contrast, though on its own, must claim the environment as source and sink, from which to abstract energy and materials and into which to excrete them. A botanical organism is partly a special kind of cause and effect system, and partly something more: partly a historical information system with a genetic coding that enables it to cope, to make a way through the world. Plants thus arise out of earthen sources (as do rocks) and turn back on their sources to make resources out of them (unlike rocks).

All this, from one perspective, is just biochemistry—the whir and buzz of organic molecules, enzymes, proteins—as humans are too from one perspective. But from an equally valid—and objective—perspective, the morphology and metabolism that the organism projects is a valued state. *Vital* is a more ample word now than *biological*. A life is spontaneously defended for what it is itself, without necessary further contributory reference, although in ecosystems such lives necessarily do have further reference. Plants defend their lives; much is valuable to them for their survival.

Plants are unified entities of the botanical though not of the zoological kind. That is, they are not unitary organisms highly integrated with centered neural control, but they are modular organisms, with a meristem that can repeatedly and indefinitely produce new vegetative modules, additional stem nodes and leaves when there is available space and resources, as well as new reproductive modules (fruits and seeds) that can organize more of that species kind. This botanical program is coded in the DNA, informational core molecules, without which the plant would collapse into the humus.

So far we have only botanical description, even when we are describing what is of value to the plant. We pass to philosophical value when we recognize that the genetic set is a *normative set*; it distinguishes between what *is* and what *ought to be*. This does not mean that the organism is a moral system, but the organism is an axiological, evaluative system. So the oak grows, reproduces, repairs its wounds, and resists death. The physical state that the organism seeks, idealized in its programmatic form, is a valued state. Every organism has a *good-of-its-kind*; it defends its own kind as a *good kind*.

The plants don't care, so why should I?, the traditional ethicists will complain. Ethics is about what people care about, that is, their values. Maybe, by extension, ethics can stretch to what animals care about. Then ethics is over.

But plants do care—using botanical standards, the only form of caring available to them. The plant life *per se* is defended. These things are not merely to be valued *for me and my kind* (as resources), not even as goods of *my kind* (sharing sentience or protein structures), but as goods *of their kind*, as *good kinds* without consideration of their kinship. So environmental philosophy, though it begins in human affairs, spreads into territories we share with neighboring organisms, such as mammals and other vertebrates. With deeper penetration, environmental philosophy evaluates all of life.[13]

When humans encounter such living organisms, they become responsible for their behavior toward them. A moral agent deciding behavior ought to take account of the consequences for other evaluative systems. We do have a responsibility to protect values, anywhere they are present and at jeopardy by our behavior. Of course, given our own biological needs, humans must eat. Humans too have to make a way through the world, and this requires capturing values present in plants and animals. Humans do so not only as biological agents but as moral agents. We have, if you like, a right to eat; we also have a responsibility to respect the vitalities of the flora around us.

4. Ethics, Endangered Species, and Biodiversity

At the species level, responsibilities increase. So does the intellectual challenge of defending duties to species. The question is partly scientific, one to be answered by the biologists. What are species? The question is partly ethical, one to be answered by the philosophers. One trouble is that scientists can find it difficult to say what a species is. Some are inclined to say that a species is merely an arbitrary classification, like the lines of latitude and longitude. Charles Darwin wrote, "I look at the term species, as one arbitrarily given for the sake of convenience to a set of individuals closely resembling each other."[14]

Indeed, biologists routinely put after a species the name of the "author" who, they say, "erected" the taxon. Sometimes it can sound like species are just decisions made by taxonomists at the universities, who make them up this way or that. They are just sets of individuals (such as bears), which can be regrouped this way and that (as bear biologists do when they dispute whether the Eurasian brown bear (*Ursus arctos*) is the same species as the North American grizzly (*U. horribilis*) and the Alaskan brown bear (*U. middendorffi*)). Nobody doubts that the individual bears exist, but if the various species are only the arbitrary groupings of biologists, one can seriously doubt whether there is a duty to endangered species.

Fortunately, biologists (including Darwin) also view species as quite real. Species are quite real as historical lineages; that there really are bear-bear-bear sequences over long periods of time is not doubted by anyone either. The species line is this reproductive process, about which there is a kind of unity and integrity, though everyone also knows that species are dynamic and changing, and can evolve into new species. G. G. Simpson concludes: "An evolutionary species is a lineage (an ancestral-descendant sequence of populations) evolving separately from others and with its own unitary evolutionary role and tendencies."[15]

Ernst Mayr holds: "Species are groups of interbreeding natural populations that are reproductively isolated from other such groups."[16] Niles Eldredge and Joel Cracraft find: "A species is a diagnosable cluster of individuals within which there is a parental pattern of ancestry and descent, beyond which there is not, and which exhibits a pattern of phylogenetic ancestry and descent among units of like kind." Species, they insist, are *"discrete entities in time as well as space."*[17] The claim that there are specific forms of life historically maintained in their environments over time does not seem arbitrary or fictitious at all but, rather, as certain as anything else we believe about the empirical world, even though at times scientists revise the theories and taxa with which they map these forms.

So species exist and are as real as individual plants or animals. The individual represents (re-presents) a species in each new generation. It is an individual ("token", as philosophers say) of a type, and the type is more important than the token. Now the philosophers can begin to ask their question, whether there can be duties to species.

But when we try to articulate this ethic, we get lost in unfamiliar territory. Natural kinds, such as species, are obscure objects of concern. Species, though they can be endangered, can't care, in the familiar senses of "care"—so comes an objection we heard before; now in a new form. They just come and go. Ninety-eight percent of the species that have inhabited Earth are extinct. It seems odd to say that species have rights, or moral standing, or need our sympathy, or that we should consider their point of view. A species lacks moral agency, reflective self-awareness, sentience, or organic individuality.

Probably most of us would say that one ought not needlessly to destroy endangered species. But many would give humanistic reasons, and think this enough. We would not say that the needless destruction of a plant species was doing something wrong to the plants, but we might say that it was vandalism in insensitive persons. Still that does not end the question, because we at once ask what are the properties in this or that endangered species, to which a person should be sensitive. Judgments of disgust and vandalism are derived from an admiration for something of value in the organisms, and if the type counts more than the tokens, then duties are to their species lineage, to the ongoing process as much as to the particular products.

When environmentalists care about endangered species, they do censure insensitivity in persons, but they also seem to appreciate an objective vitality in the world, one that precedes and overleaps human personal or cultural preferences. To care about endangered species, then, is not to report some subjective preferences in humans who fancy rare plants or animals. To the contrary, it is to be quite nonanthropocentric and objective about botanical and zoological processes that take place independently of human preferences.

In species, there is a biological identity reasserted genetically over time. The life that the individual has is something passing through the individual as much as something it intrinsically possesses, and a respect for life finds it appropriate to attach duty dynamically to the specific form of life. The species line is the dynamic living system, the whole, of which individual organisms are the essential parts. The species too has its integrity, its individuality, its "right to life" (if one chooses to use the rhetoric of rights); and it is more important to protect this vitality than to protect individual integrity. The right to life, biologically speaking, is an adaptive fit that is right for life, that survives over millennia, and this generates at least a presumption that species are good and therefore that it is right for humans to let them be, to let them evolve. The appropriate survival unit is the appropriate level of moral concern.

A shutdown of the life stream on Earth is the most destructive event possible. In threatening Earth's biodiversity, the wrong that humans are doing, or allowing to happen through carelessness, is stopping the historical vitality of life. Every extinction is an incremental decay in this stopping of life, no small thing. "Ought species x to exist?" is a distributive increment in the collective question, "Ought life on Earth to exist?" Since life on Earth is an aggregate of many species, when humans jeopardize species, the burden of proof lies with those who wish deliberately to extinguish a species and simultaneously to care for life on Earth.

One form of life has never endangered so many others. Never before has this level of question been deliberately faced. Humans have more understanding than ever of the natural world they inhabit, of the speciating processes, more predictive power to foresee the intended and unintended results of their actions, and more power to reverse the undesirable consequences. At this point, all biology ought to become conservation biology, committed to optimizing the values carried by species. Any philosopher, examining life (as Socrates urged), ought to see that the responsibilities that such power and vision generate no longer attach simply to individuals or persons but are emerging duties to specific forms of life. What is required is principled responsibility to the biospheric Earth.

Few philosophers in the classical past have ever raised the question of duties to species, much less answered it. But now such duty is becoming clearer. Indeed it is urgent. If, in this world of uncertain moral convictions, it makes any sense to claim that one ought not to kill individuals without justification (as philosophers have said since Socrates), it makes more sense to claim that one ought not to kill the species without extraordinary justification. Several billion years worth of creative toil and several million species of teeming life have been handed over to the care of this late-coming species in which mind has flowered and morals have emerged. Life on Earth is a many splendored thing; extinction dims its luster.

From here onward, no one can claim to be living an examined life, to be examining life on Earth, unless he or she knows this responsibility to species and acts accordingly. Were the eminent moral species, *Homo sapiens*, to conserve all Earth's species merely as resources for human preference satisfaction, we would not yet know the saving truth about what is or ought to be going on in biological conservation. If you believe that, you are already doing environmental philosophy. If you do not, here is an invitation to start.

5. Ecosystems, a Land Ethic, and Ethics in Place

We have been traveling into progressively less familiar ethical terrain, though biologically quite fundamental. Ecosystems are ultimately—at least on the earthen scene—our home, from which *ecol-*

ogy is derived (Greek: *oikos*, house). We need a logic and an ethic for Earth with its family of life.

"A thing is right," urged Aldo Leopold, "when it tends to preserve the integrity, stability, and integrity of the biotic community; it is wrong when it tends otherwise."[18] "That land is a community is the basic concept of ecology, but that land is to be loved and respected is an extension of ethics." "When we see land as a community to which we belong, we may begin to use it with love and respect."[19] "The land ethic simply enlarges the boundaries of the community to include soils, waters, plants, and animals, or collectively: the land."[20] Ethics is here by people but not just for people; one needs an ethic of place.

But ecosystems are unfamiliar moral territory; it is difficult to get the biology right (again, as with species before), and, superimposed on the biology, difficult to get the ethics right. Fortunately, it is often evident that human welfare depends on ecosystemic support, and in this sense even those who believe that ethics is only about people can support legislation about clean air, clean water, soil conservation, forest policy, pollution controls, renewable resources and so forth, which deals with ecosystem-level processes. Further, humans find much of value in preserving wild ecosystems, for instance in our wilderness and park systems and our biological reserves. Still, a comprehensive environmental ethics needs the best, naturalistic reasons, as well as the good, humanistic ones, for respective ecosystems.

Again, we have a scientific question mixed with an ethical one. What are ecosystems? Only after answering that question can we ask the full extent of value present there, and whether humans can have duties to ecosystems. We need an accurate description of ecosystems and an informed prescription for conduct. We have to make clear, both in science and in ethics, a paradigm of community. Earlier we heard John Passmore claim that although ecosystems might be biotic communities, they are not moral communities. He is right that the members who are not humans are not reciprocating moral agents. But is he right that ecosystems cannot count morally?

Ecologists themselves have had differing opinions about ecosystems. The debate among the biologists has, understandably, con-

fused the philosophers. To some, ecosystems have seemed to be little more than stochastic (probabilistic, random) processes. A
seashore, a tundra is a loose collection of externally related parts.
Much of the environment is not organic at all (rain, groundwater,
rocks, nonbiotic soil particles, air). Some is dead and decaying de-
bris (fallen trees, scat, humus). These things have no organized
needs; the collection of them is a jumble, hardly a community.
Though the plants and animals within an ecosystem do have
needs—each defends its own life—the fortuitous interplay between organisms is simply a matter of the distribution and abundance of organisms, how they get dispersed here and not there,
birthrates and deathrates, population densities, moisture regimes,
parasitism and predation, checks and balances. There is really not
enough centered process to call community. There is only catch-
as-catch-can scramble for nutrients and energy.

We respect a plant or animal because every organism defends an
"organized" biological identity. An ecosystem is the necessary
habitat for this, but an ecosystem itself has no genome, no brain,
no self-identification. It does not defend itself against injury or
death as do bears or even daisies. It is not irritable. An oak-hick-
ory forest has no telos, no unified program it is set to execute. The
parts (bears and daisies) are more complex, some will say, than the
wholes (forests, grasslands).

So it can begin to seem as if concern for ecosystem is secondary
after all, instrumental to a respect for human and nonhuman life.
An ecosystem is too low a level of organization to be the direct focus of concern. Ecosystems have no interests about which they or
we can care. More troublesome still, an ecosystem is a place of contest and conflict, a jungle where the fittest survive, beside which
organisms are models of integrated cooperation, and animals are
the centers of psychological experience. The so-called "community" is pushing and hauling between rivals, or indifference and
haphazard juxtaposition, nothing calling forth our admiration.

But to say that and nothing more is to misunderstand ecosystems. Painting a new picture on the conflict side, even before the
rise of ecology, biologists concluded that to portray a gladiatorial
survival of the fittest was a distorted account. They prefer a model

of the better adapted fit. Although conflict is part of the picture, the organism is selected for a situated environmental fitness. A bear fits a forest just as much as its heart fits its lungs. There are differences; the heart and lungs are close-coupled in a way that bear and forest are not. Still, the bear requires its forest community; the bear-organism fits there, as surely as its organs fit together to compose a bear.

There is a crucial element of struggle, but it is equally important to see this struggle contained in community. Ecological science emphasizes how there is a biological sense in which the integrity, beauty, and stability of each individual and species is bound up with their coactions. Predator and prey, parasite and host, grazer and grazed require a coevolution where both flourish, since the health of the predator, parasite, grazer is locked into the continuing existence, even the welfare, of the prey, host, or grazed. Ecosystems are not of disvalue because contending forces are in dynamic process there, any more than cultures are. Like business, politics, and sports, ecosystems thrive on competition.

The community connections, though requiring adaptive fit, are more loose than the organismic coactions. But that does not mean they are less significant. Concentrated unity is admirable in the organism, but the requisite matrix of its generation is the open, plural ecology. Internal complexity arises to deal with a complex, tricky environment. The skin–out processes are not just the support, they are the subtle source of the skin–in processes. Had there been either simplicity or lock-step concentrated unity in the surroundings, no creative unity could have been composed internally. There would have been less elegance in life.

To look to ecosystems for what we respect in individual animals and plants, and to find such characteristics missing, and then judge that ecosystems do not count morally, makes what philosophers call a category mistake. To look at one level for what is appropriate at another faults *communities* as though they ought to be organismic *individuals.* One should look for a matrix of interconnections between centers, not for a single center, for creative stimulus and open-ended potential, not for a fixed telos and executive program. Everything will be connected to many other things, sometimes by obligate associations, more often by partial and pliable dependencies; and, among other components, there will be no significant interactions. There will be shunts and criss-crossing pathways, cybernetic subsystems and feedback loops, functions in a communal sense. One looks for selection pressures and adaptive fit, for speciation and life support.

An ecosystem systematically generates a spontaneous order that exceeds in richness, beauty, integrity, and dynamic stability the order of any of the component parts, an order that feeds (and is fed by) the richness, beauty, and integrity of these component parts. Though these organized interdependencies are "loose" in comparison with the "tight" connections within an organism, all these metabolisms are vitally linked. The equilibrating ecosystem is not merely push-pull forces. It is an equilibrating of values. Ecosystems select for adaptive fit; they have generated over evolutionary time increasingly richer lives in quality and quantity, and continue now to support myriads of species and individuals, with higher levels of autonomy and experience at the top trophic levels.

Using criteria appropriate to this level, philosophers ought to find that such ecosystems are satisfactory communities to which to attach duty. Our concern must be for the fundamental unit of survival. Ecosystems are the womb of life, the home community. Human cultures emerge from Earth's ecosystems and remain tethered to them. If such biotic communities are not admirable, satisfactory, and morally considerable, why not?

On the humanistic account, such species in their ecosystems ought to be saved for their benefits to humans. On the naturalistic account, the sole moral species has a duty to do something less self-interested than count all the products of an evolutionary ecosystem as nothing but human resources. Rather, the host of species and the system producing them has a claim to care in its own right. There is something Newtonian, not yet Einsteinian—besides something morally naive, also perhaps myopic and arrogant—about living in a reference frame where one species takes itself as absolute and values everything else relative only to its utility.

5. Earth Ethics

There is a sense in which the term "land ethic," chosen by Aldo Leopold when he urged expanding ethics in the Wisconsin sand counties, is a little unfortunate. He was right enough about the need where he was, right enough about where to start, and he did launch a seminal invitation to environmental philosophy. But in the half century since, we have had to enlarge ethics to include global concerns, to cover the Earth. Leopold's "land" is really the biotic community of life, because he included the rivers and the soils, as well as the fauna and flora. Still, his vision is local. A "land" ethic is not marine, for example. It does not ask about global warming, or ozone holes. Leopold did not ask about the population explosion or about sustainable development, not at least about the rich developed nations and the poor developing ones. Today, the horizons are expanding. We have got to look the whole Earth over to get ethics really in place.

Boutros Boutros-Ghali, speaking as the UN Secretary-General, closed the Earth Summit: "The Spirit of Rio must create a new mode of civic conduct. It is not enough for man to love his neighbour; he must also learn to love his world."[21] "We must now conclude an ethical and political contract with nature, with this Earth to which we owe our very existence and which gives us life."[22] This does not deny that we must continue to love our neighbors, but it enlarges the vision from just a social contract to a natural contract. The challenge is to think of Earth as a precious thing in itself because it is home for us all; Earth is to be loved, as we do a neighbor, for an intrinsic integrity. The center of focus is not people, but the biosphere. That is the reformation that the earthstruck astronauts with their whole Earth photographs invite.

Even more than valuing this or that particular ecosystem, and finding duties in result, valuing the whole Earth and responsibilities to it are unfamiliar and need philosophical analysis. A duty to the planet? Be careful, the hard-nosed humanist philosophers will say. Keep your logic in reign. Earth is really just a big rockpile like the moon, only one on which the rocks are watered and illuminated in such a way that they support life. We cannot have duties to dirt. Or to oceans, or mountains. So it is really the life we value and not the Earth, except as instrumental to life. We have duties to people, perhaps to animals that can suffer pains and pleasures, somewhat less plausibly to all living things. But we must not confuse duties to the home with duties to the inhabitants. We do not praise the earth so much as what is on Earth.

But this is not a systemic view of the valuable Earth we now behold, before we beheld it. Overlooking Earth, we confront not just some value that is generated in the eye of the beholder. Or some value that is found in this or that creature on it. Finding this more comprehensive level of value will generate a global sense of obligation. That people might have duties to dirt is often taken to be the *reductio ad absurdum* in philosophy. Philosophers often defeat an argument by showing that it leads logically to an absurdity. Those doubtful of environmental philosophy may think that once the idea of duties to others than humans starts, one slides down a slippery slope—animals, plants, species, ecosystems, mountains, rivers, clouds, dirt—and ends up claiming the ridiculous: that rocks have rights.

But this depends on how much "dirt" one considers. Shift the focus from *e*arth to *E*arth. The evolution of rocks into dirt into fauna and flora is one of the great surprises of natural history, one of the rarest events in the astronomical universe. We humans too arise up from the humus, and we find revealed what earth can do when it is self-organizing under suitable conditions. This is pretty spectacular dirt. On an everyday scale earth seems to be passive, inert, an unsuitable object of moral concern. But on a global scale?

The scale changes nothing, a critic may protest, the changes are only quantitative. Earth is no doubt precious as life support, but it is not precious in itself. There is nobody there in a planet. There is not even the objective vitality of an organism, or the genetic transmission of a species line. Earth is not even an ecosystem, strictly speaking; it is a loose collection of myriads of ecosystems. So we must be talking loosely, perhaps poetically, or romantically of valuing Earth. Earth is a mere thing, a big thing, a special thing for those who happen to live on it, but still a thing, and not appropriate as an object of intrinsic or systemic valuation. Thinking this way, we can, if we insist on being anthropocentrists, say that it is all valueless except as our human resource.

But we will not be valuing Earth objectively until we appreciate this marvelous natural history. This really is a superb planet, the most valuable entity of all, because it is the entity able to produce all the Earthbound values. At this scale of vision, if we ask what is principally to be valued, the value of life arising as a creative process on Earth seems a better description and a more comprehensive category than to speak of a careful management of planetary natural resources that we humans own. Such a fertile Earth, interestingly, is the original meaning of the word "nature," that which "springs forth," or "gives birth," or is "generated." This was once explained in the mythology of a "Mother Earth"; now we have it on scientific authority.

Dealing with an acre or two of real estate, perhaps even with hundreds or thousands of acres, we can think that the earth belongs to us, as private property holders. Dealing with a landscape, we can think that the earth belongs to us, as citizens of the country geographically located there. But on the global scale, Earth is not something we own. Earth does not belong to us; rather we belong to it. We belong on it. The deeper philosophical question is how we humans belong in this world, not how much of it belongs to us. The latter is only an economic question. This is an invitation to environmental philosophy. The question is not of property, but of community. The valuing of nature is not over until we have risen to the planetary level, and valued this system we inhabit. Earth is really the relevant survival unit. And with that global vision, we may want to return to our regional landscapes, such as the sand counties of Wisconsin, and think of ourselves as belonging there too, with a deeper sense of place.

7. Ethics for a New Millennium

Until recently, the mark of an educated person could be summed up as *civitas*, citizenship. People ought to be productive in their communities, leaders in business, the professions, government, church, education. That was what colleges and universities tried to produce: educated citizens. That was why people studied philosophy: to examine the good life that humans can and ought to choose for themselves, their highest good, what philosophers called the *summum bonum.* The mark of an educated person is today, increasingly, something more. Ethics is becoming something different, as we turn a century.

In the next millennium, as we already realize, it will not be enough to be a good "citizen," or a "humanist," because neither of those terms have enough "nature," enough "earthiness" in them. "Citizen" is only half the truth; the other half is that we are "residents" on landscapes. Humans are Earthlings. Earth is our dwelling place. From here onward, there is no such thing as civic competence without ecological competence. Many a citizen who is celebrated for his or her humanity is quite illiterate when it comes to reading the signs of the times boding ecological crisis, or, even were there no crisis, in enjoying the values the natural world carries all around them. Philosophy professors stimulate their students to think about their duties to fellow citizens; this is commendable, but, alas, such teachers can leave their students without a sense of responsibility on their native landscapes. Neither professor nor students yet have a land ethic, an Earth ethics. Until that happens, no one is well educated for the next century, the century in which many of these problems will have to be solved, if ever they are solved.

Our responsibility to Earth might be thought the most remote of our responsibilities; it seems so grandiose and vague beside our concrete responsibilities to our children or next-door neighbors. But not so: the other way around, it is the most fundamental of our responsibilities, and connected with these local ones. Responsibilities increase proportionately to the level and value of the reality in jeopardy. The highest level that we humans have power to affect, Earth is the most real phenomenon of all, marvelously real. We can hardly be responsible to anything more cosmic, unless perhaps to God.

Real community does not yet exist at world levels; nevertheless humans live on only one Earth and our powers operate at global ranges. An opportunity that we face from here onward, indeed a necessity thrust upon us, is to see Earth globally, to see ourselves as Earth residents with transnational interests. From the perspective of a nation state, when we hear the word "international," we think at once of domestic and foreign. But with the word "global," there is no domestic and foreign, we are all natives. At that level, we are not citizens of a nation but "residents." We like to think of ourselves as "cosmopolitan," appreciating our own culture, as well as the accomplishments of many other cultures. But the animal who builds a "polis" still inhabits an "oikos," a whole world; humans have an ecology. We are incarnate in earth; we are Earth incarnate.

The natural and the cultural on Earth have entwined destinies. Across great reaches of geological time, there were no humans on Earth; Earth was entirely a natural system. Earth remains a vast natural system, as we see with the views from space. But for several thousand years Earth has increasingly supported cultural systems, and, in the last few centuries, these cultural systems have exploded. The great universities of the world have fueled that explosion, providing the knowledge that has made the modern world possible, with its vastly developed economies. Today, everywhere, the resulting explosion of culture presses Earth's natural systems to their carrying capacities.

Diverse combinations of nature and culture have worked well enough over many millennia, but no more. Our modern cultures threaten the stability, beauty, and integrity of Earth, and thereby of the cultures superposed on Earth. An interhuman ethics must serve to find a satisfactory fit for humans in their communities; and, beyond that, an environmental ethics must serve to find a satisfactory fit for humans in the larger communities of life. We worried throughout much of this century that humans would destroy themselves in interhuman conflict; that fear has subsided somewhat only to be replaced by a new one. The worry for the next century is that, if our present heading is uncorrected, humans may ruin their planet and themselves with it.

Colleges and universities are supposed to defend the arts and the sciences, the wisdom of the human genius. Colleges and universities guard our humanity. They transmit the heritage of culture, without which we cannot be human. They teach humans the art of living well; they teach the sciences by which we understand the world and benefit from it. Animals have neither arts nor sciences; the particular virtue of *Homo sapiens*, the wise species, is this transmissible knowledge, this wisdom by which human life continues and flourishes. So it will first be thought, in keeping with this classical vision, that the role of the university is to protect human values at stake.

Now along comes environmental philosophy inviting a radically different inquiry. Environmental philosophy asks whether or not the colleges and universities are not themselves part of the problem, as much as part of the answer. They have produced the knowledge by which humans have gained their startling powers for the rebuilding and the degradation of this home planet. The knowledge accumulated in the universities, transmitted from one generation to the next, is of great genius. Yet it has destabilized human life on our home planet. Both the sciences and the humanities are responsible. Business too is responsible; so is politics, and religion. But our concern here is that philosophy and ethics are responsible, and need to figure in a better education. Universities set the pace intellectually; they educate today the leaders of tomorrow. In that sense, the challenges of environmental ethics are challenges to liberal education in the arts and sciences.

Socrates said that he was a lover of wisdom, and we admire him for that. But he once added, "I'm a lover of learning, and trees and country places won't teach me anything, whereas people in the city do."[23] Socrates loved the city with its academy, its politics and culture, but avoided nature as profitless and boring. On the other hand, when John Muir finished his formal education and turned to live in the Sierra Nevadas, he wrote, "I was only leaving one university for another, the Wisconsin University for the University of the Wilderness."[24] Colleges and universities love learning; they love people and strive to make and keep life human. But, in an environmental ethics, life cannot be made and kept human unless we know our place, the human residence on this home planet.

We think of the universities as being the scene of an explosion of knowledge over the decades of this century, and we hope for more in the century to come. We hope such learning is endless in the new millennium. We think that knowledge is power. But if this explosion of knowledge and its resulting empowerment has produced a planet in crisis, perhaps the genius of the university is not what we thought. Our knowledge has not in every respect made us better fitted for life on the planet; in important respects it has made us misfits. We could even be, or become, so misfitted that our welfare, or even our survival, is at stake.

Evolutionary history has been going on for billions of years, while cultural history is only about a hundred thousand years old. But certainly from here onward, culture increasingly determines what natural history shall continue. In that sense, it is true that Earth is now in a post-evolutionary phase. Culture is the principal determinant of Earth's future, more than nature; we are passing into a century when this will be increasingly obvious. The next millennium, some are even saying, is the epoch of the end of nature.

Environmental philosophy invites another vision, the inquiry whether we humans can launch a millennium of culture in harmony with nature. After all, the technosphere remains in the biosphere; we are not in a post-ecological phase. The management of the planet must conserve some environmental values, if only for our survival, and it ought to conserve many more, if we are to be wise. Hopefully, such policy can, in places, let nature take its course, let wild nature be. Here is the challenge John Muir found in his University of the Wilderness:

> The world, we are told, was made especially for man—a presumption not supported by all the facts. A numerous class of men are painfully astonished whenever they find anything, living or dead, in all God's universe, which they cannot eat or render in some way what they call useful to themselves. . . . Now, it never seems to occur to these far-seeing teachers that Nature's object in making plants and animals might possibly be first of all the happiness of each of them, not the creation of all for the happiness of one. Why should man value himself as more than a small part of the one great unit of creation?[25]

Earth is a fragile planet, a jewel set in mystery. We humans too belong on the planet; it is our home, as much as for all the others. Like it or not, we have a dominion here; we do have to learn to handle ourselves and to manage our cultures and the supporting environment. We can and ought to examine our lives and our world and choose the highest goods. We are, in that sense, more than just another small part of creation. We have a big responsibility, that of discovering an appropriate respect for nature.

Environmental ethics, you may have thought at the start, is marginal, an ethic for the chipmunks and daisies, for canoe freaks and tree huggers. Not so, environmental ethics is right at the center of the challenges of the next millennium.

Notes

1. Astronomer Fred Hoyle, quoted in Kevin W. Kelley, ed., *The Home Planet* (Reading, MA: Addison-Wesley, 1988), inside front cover.

2. Quoted in Kelley, at photographs 42–45.

3. Michael Collins, "Foreword," in Roy A. Gallant, *Our Universe* (Washington, DC: National Geographic Society, 1980), p. 6.

4. John Passmore, *Man's Responsibility for Nature* (New York: Scribner, 1974), p. 116.

5. Bryan G. Norton, *Toward Unity Among Environmentalists* (New York: Oxford University Press, 1991) claims that fully–enlightened anthropocentrists and more naturalistic environmentalists will almost entirely agree on environmental policy, what he calls a "convergence hypothesis."

6. Susan S. Hanna, Carl Folke, and Karl-Göran Mäler, eds., *Rights to Nature* (Washington, DC: Island Press, 1996).

7. *Environmental Protection and Sustainable Development: Legal Principles and Recommendations* (London/Dordrecht, Netherlands: Graham and Trotman/Martinus Nijhoff Publishers, 1987), p. 9.

8. Lee Hannah, David Lohse, Charles Hutchinson, John L. Carr and Ali Lankerani, "A Preliminary Inventory of Human Disturbance of World Ecosystems," *Ambio* 23 (1994), pp. 246–50.

9. Mary Midgley, *Animals and Why They Matter* (Athens, GA: University of Georgia Press, 1983).

10. Mary-Claire King and A. C. Wilson, "Evolution at Two Levels in Humans and Chimpanzees," *Science* 188 (1975), pp. 107–16. This is, however, only for structural genes, not for genes that regulate behavior.

11. Aldo Leopold, *A Sand County Almanac* (New York: Oxford University Press, 1968), p. 130.

12. Martin J. Wells, *Octopus* (London: Chapman and Hall, 1978), pp. 8–9.

13. Paul W. Taylor, *Respect for Life* (Princeton, NJ: Princeton University Press, 1986) calls this "biocentrism." Albert Schweitzer earlier spoke of a "reverence for life."

14. Charles Darwin, *The Origin of Species* (Baltimore, MD: Penguin Books, 1968), p. 108.

15. G. G. Simpson, *Principles of Animal Taxonomy* (New York: Columbia University Press, 1961), p. 153.

16. Ernst Mayr, *Principles of Systematic Zoology* (New York: McGraw-Hill, 1969), p. 26.

17. Niles Eldredge and Joel Cracraft, *Phylogenetic Patterns and the Evolutionary Process* (New York: Columbia University Press, 1980), p. 92.

18. Leopold, *Sand County Almanac*, pp. 224–25. Leopold's principal philosophical interpreter is J. Baird Callicott, *In Defense of the Land Ethic* (Albany, NY: State University of New York Press, 1989).

19. Ibid., pp. viii–ix.

20. Ibid., p. 204.

21. Boutros Boutros-Ghalli, Extracts from closing UNCED statement, in an UNCED summary, *Final Meeting and Round-up of Conference*, June 14, 1992, p. 1. UN Document ENV/DEV/RIO/29, 14 June 1992.

22. Boutros Boutros-Ghalli, Text of closing UNCED statements, in *Report of the United Nations Conference on Environment and Development*, 1992, vol. IV, pp. 66–69. UN Document A/CONF.151.26 (Vol. IV).

23. Plato, *Phaedrus* 230d., trans. H. H. Fowler, *Plato*, vol. 1, Loeb Classical Library (Cambridge, MA: Harvard University Press, 1914, 1977), pp. 423–24.

24. John Muir, *The Story of My Boyhood and Youth* (Madison: University of Wisconsin Press, 1965), p. 228.

25. John Muir, *A Thousand Mile Walk to the Gulf* (Boston: Houghton Mifflin Co., 1916), pp. 136–39.

5

The Journey Home

Jim Cheney

Mindfulness and Knowledge

Missing in modern conceptions of knowledge is a sense of active
and reciprocal communication with the nonhuman world. On an
older understanding, knowledge emerges from a conversation between world and person, and our human part in the genesis of

knowledge, in its most essential aspect, is to prepare ourselves ethically and spiritually for the reception of knowledge.

Modern conceptions of knowledge, by contrast, hold that knowledge emerges from a relationship between an active, knowing subject and a passive, known object. Knowledge is a monologue, not a conversation. It is forged and maintained by unifying and controlling human unconsciousness; it is lodged in the human mind and represents the world as human activity and human interests have determined it to be.

The anthropologist Henry Sharp says of a Chipewyan Indian conception of knowledge that

> Symbols, ideas, and language . . . are not passive ways of perceiving a determined positivist reality but a mode of interaction shared between the *dene* and their environment. All animate life interacts and, to a greater or lesser degree, affects the life and behavior of all other animate forms. In their deliberate and splendid isolation, the Chipewyan interact with all life in accordance with their understanding, and the animate universe responds.

Note carefully: Sharp says that language is a *mode of interaction*, not merely a symbol system for describing the world. As a mode of interaction that *affects the life and behavior of all other animate forms*, our use of language can be seen to have an ethical dimension. The mindfulness—or etiquette—we bring to our interaction with the world shapes our knowledge of that world.

And, because it shapes our knowledge, the etiquette we bring to our attempts to understand the world shapes the world itself, as Sharp's comments on White Canada's interaction with the nonhuman world indicate:

> White Canada does not come silently and openly into the bush in search of understanding or communion, it sojourns briefly in the full glory of its colonial power to exploit and regulate all animate being and foremost of all, the *dene*. It comes asserting a clashing causal certainty in the fundamentalist exercise of the power of its belief. It talks too loudly, its posture is

wrong, its movement harsh and graceless; it does not know what to see and it hears nothing. Its presence brings a stunning confusion heard deafeningly in a growing circle of silence created by a confused and disordered animate universe.[1]

Sharp believes that the Chipewyan understand their relationship to the nonhuman world as a reciprocal relationship to active moral beings, sources of knowledge and power. Knowledge in such a world clearly has an ethical dimension: the quest for knowledge is conducted in accordance with a particular etiquette, and the world responds accordingly, shaping knowledge in turn.

The dominant Western conception of knowledge—exemplified by "White Canada"—is also governed by an etiquette, though in this case an "etiquette" shaped by the values of domination and control. These values determine objects of knowledge as passive, mere objects of study and manipulation, not agents in the shaping of knowledge.

Can we accept the view that the nonhuman world is an active participant in the construction of knowledge? A noted mathematical biophysicist answers "yes":

> "Laws" of nature name nature as blind, obedient, and simple; simultaneously, they name [the scientist] as authoritative, generative, resourceful, and complex. . . . By contrast, the conception of nature as orderly, and not merely law bound, allows nature itself to be generative and resourceful—more complex and abundant than we can either describe or prescribe. In this alternative view, nature comes to be seen as an active partner in a more reciprocal relation to an observer. . . . Such a relationship . . . would require a different style of inquiry, no less rigorous but presupposing the modesty and open attentiveness that allow one to "listen to the material.". . .[2]

The earth itself as active partner in the construction of knowledge! This conception of knowledge is what I understand by a spiritual relationship to the earth. As another writer adds, however, giving up mastery in our search for knowledge also makes room "for some unsettling possibilities, including a sense of the world's independent sense of humor." American Indian tricksters embody these unsettling possibilities.[3]

Considering Rocks

Nietzsche says that "All experiences are moral experiences, even in the realm of sense perception."[4] Imagine a deep practice of universal consideration[5] and considerateness that is not so much instituted as a principle or rule governing one's behavior as it is a dimension of one's very *perception* of the world—and I do think that this requires an act of considerable moral imagination for those of us raised in the heart of the monster, the Western dualism of moral insiders and outsiders, a rent in the fabric of the universe that is part and parcel of an understanding of the world governed by values of domination and control.

Such a conception is present in the notion of "respect" for other-than-human nature that is so pervasive in Indigenous cultures. To Western ears, this term may have overtones of hierarchically structured relationships, or it might sound like obedience to moral law. But to Indigenous ears it points to a mode of presence to the world the central feature of which is *awareness* of all that is, an awareness that is simultaneously a mode of knowing and an etiquette, as Carol Geddes explained in response to a question concerning the meaning of the Indigenous notion of respect: "it does not have a very precise definition in translation—the way it is used in English. It is more like awareness. It is more like knowledge and that is a very important distinction, because it is not like a moral law, it is more like something that is just a part of your whole awareness. It is not something that is abstract at all."[6]

Consider rocks. That is, consider rocks as a case study of a mode of understanding informed by the deep practice of ethical consideration implied by the Indigenous notion of "respect." Although the very idea that we might have ethical relationships with *rocks* is nearly unintelligible to the Western mind (or proof of mental imbalance), many American Indian cultures take rocks quite seriously, often regarding Rock as the oldest and (in some sense) wisest of beings.

Suppose we begin with the idea that knowledge arises as a "knowing *with*" the earth, that the earth (rocks included) is co-participant in the shaping of knowledge and that this relationship has an ethical dimension, is governed by an etiquette. The shape and content of such knowledge will, of course, change as one switches from a mode of knowledge governed by values of domination and control (and its attendant etiquette) to a mode of knowledge (and its attendant etiquette) that opens one to the possibility that rocks, by their very existence and presence to us, might give rise to a different order of understanding.

If we do this, we might come to understand that existence itself is sacred, more-than-human; that is, that the particular virtues of specifically human being are embedded in mystery, embedded in, and nourished by, a broader, deeper, more powerful and enduring matrix.

How might rocks teach us this? Perhaps by their very presence: they are ancient, enduring presences, the oldest of beings. They are, perhaps, "watchful." (Here I start using scare quotes. But the use of metaphor here is not careless writing. Knowledge moves by metaphor. We must, of course, be careful, critical, and attentive in our use of metaphors—that they may reach *insightfully* into mystery.) An important aspect of any learning situation is mindful presence. In this, rocks, in their enduring presence, their watchfulness, may be our first and most profound teachers of the most fundamental aspects of moral presence in and to this world: mindfulness and universal consideration, universal invitation into the reciprocities of knowledge and care.

Something closely related to this is expressed by the character Stone Columbus in Gerald Vizenor's *The Heirs of Columbus*: "Stones hold our tribal origins and our worlds in silence in the same way that we listen in stories and hold our past in memories."[7] And Lee Irwin reports that "the sacred stone . . . is prominent in the [medicine] bundles of all the Plains people. . . . Generally, its symbolism is tied to an ancient knowledge that the earth alone possesses."[8]

What is it to treat rocks with ethical regard?

We might begin by considering certain facts about rocks with a kind of ethical mindfulness. Rich, brown billion-year-old volcanic basalt—so strongly present to me during this year away from

my home, writing this chapter and working with the Native Philosophy Project in Thunder Bay—overlies the even more ancient
two and three billion-year-old granite of the Canadian Shield. This
ancient Precambrian rock connects me, during my year on the
north shore of Lake Superior, with my birth place in southeastern
Minnesota. I dream my way back to my birth along ancient flows
of basalt and the deep granitic basement of the north country. The
sedimentary rock of my home in southeastern Wisconsin is much
younger and is quite different in its character. (The word "character" was chosen deliberately because it is already at home for us in
both physical and personal descriptions, and helps move rocks into
the ethical domain, a trick of metaphor and ambiguity.) Sedimentary rock calls up in our minds ancient seas, ancient life. This rock
is a visible repository of those seas, that life. Canadian Shield rock
and the overlying basalt flows are different. They are born of great
pressure and the sustaining heat underlying the earth's crust. They
speak to us of this pressure and sustaining heat. They are ancient,
watchful. Their presence differs from that of sedimentary rock.

Near a high mountain lake a number of years back, a backpacking friend came upon a huge granite rock that expressed the more-
than-human nature of the world in a particularly concentrated way
that he could only refer to—with a certain awe—as "sacred." It
wasn't in any way obvious where this rock could have come from,
and yet there it was, watchful, still, present. It appeared to have
been from time out of mind in that place, by that wild and beautiful spangled lake. A still point of the turning world, an eternal,
serene, and powerful presence. Hiking back to our campsite, we began noticing the watchful presence of small stones and pebbles
along the trail. They were connected in our minds to the sacred
presence of Rock (*Inyan*, in Lakota) by that spangled lake and shared
in its sacredness. They were closer to human dimensions, both in
physical size and in character. They were, in a sense, companions,
partners, some of them quite active, youthful, with funny stories to
tell, perhaps, if we had listened with more care.

I do not want to stress these characterizations too much. They
constitute only a few tentative directions for the development of
ethical relationships to rocks. Descriptions and metaphors would
differ from person to person and from culture to culture, of course,

and that is as it should be, particularly at this originary stage of Western ethical relationships to rocks. Where this all might lead is hard to say. What I suggest, however, is that these descriptions and metaphors do at least hint at ethical practices and relationships appropriate to rocks of various kinds and characters.

What rocks teach us is experientially bound up with (is the other face of) what we come to understand to be our ethical relationships to rocks. As these relationships deepen, so do the teachings.

Ceremonial Worlds

Is there a way for us in the West to genuinely participate in such a relationship with the world? For many of us in the "academic West" it is probably necessary to find some *conceptual* trail that leads to the borders of Indian Country. The remainder of this chapter follows one such trail.

Whereas "sight presents surfaces," Walter Ong says, "sound reveals interiors" and "signals the present use of power, since sound must be in active production in order to exist at all." For this reason, voice is "the paradigm of all sound, and to it all sounds tends to be assimilated." Further, in a sound-oriented culture, "the universe [is] something one respond[s] to, as to a voice, not something merely to be inspected."[9]

Ong and others have linked the visual metaphor of knowing—as in the term "world view," used to refer to a people's fundamental beliefs about the world—with the advent of the written word. Words on the page no longer reveal interiors as do sounds, they no longer signal the present use or manifestation of power. With literacy words become *objects* visually displayed on the page; they are inert, in themselves lifeless. They become signs, symbols of something else. Sam Gill reports (as have many others) that nonliterate people are often highly critical of writing. He says of this, however, that he does

> not believe that it is actually writing that is at the core of their criticism. The concern is with certain dimensions of behavior and modes of thought that writing tends to facilitate and encourage. And these dimensions are linked to the critical, se-

mantical, encoding aspects of language. . . . We interpret texts
to discern systems of thought and belief, propositional or historical contents, messages communicated. Put more generally,
we seek the information in the text. We tend to emphasize code
at the expense of behavior, message at the expense of the performance and usage contexts.[10]

The written word conspires with the visual metaphor of knowledge to turn the world into a passive object for human knowledge
and to focus our attention on language as a sign system primarily
designed to encode belief or to represent the world by encoding
beliefs about the world.

In a number of articles in his *Native American Religious Action*,
Sam Gill has attempted to reinstate the fundamental nature of the
performative function of language, using Navajo prayer as a case
study. He points out that the response he invariably receives when
he asks Navajo elders what prayers *mean* is that they tell him "not
what messages prayers carry, but what prayers *do*." Further, "the
person of knowledge in Navajo tradition holds that "['theology,
philosophy, and doctrine'] are ordinarily to be discouraged. Such
concerns are commonly understood by Navajos as evidence that one
totally misunderstands the nature of Navajo religious traditions."[11]

Generalizing from his analysis of prayer acts, Gill asserts that
"the importance of religion as it is practiced by the great body of
religious persons for whom religion is a way of life [is] a way of
creating, discovering, and communicating worlds of meaning
largely through ordinary and common actions and behavior."[12]

I would like to explore the value for understanding so-called
world views of generalizing even further, arguing that the performative dimension of language be understood as fundamental—not
just in obviously ceremonial, ritual, or religious settings, but *generally*. We *do* things with words. Foremost among these performative functions is the creation of what I will call the *ceremonial worlds*
within which we live. Other performative functions of language
are possible only within these ceremonial worlds—promise making, for example, is possible only within an accepted set of social
conventions, as is the progress achieved within science.

Diamond Jenness reports an unnamed Carrier Indian of the Bulkley River as saying: "The white man writes everything down in a book so that it might not be forgotten; but our ancestors married the animals, learned their ways, and passed on the knowledge from one generation to another."[13] I would suggest that we understand him as saying that his people's ancestors passed down the means of creating, or recreating, the worlds, the ceremonial worlds, within which they lived—the stories, the ceremonies, the rituals, the daily practices. They passed down modes of action, which when written down come to be understood as information. The white man in this account wants to know what beliefs are encoded in the utterances of Indians, he wants to treat these utterances as mirrors of Indian worlds. But the utterances function primarily to *produce* these worlds. The white man is concerned with correct descriptions of Indian worlds. Indians in this account, on the other hand, are concerned with right relationship to those beings that populate the ceremonial world, they are concerned with mindfulness and grace. These ceremonial worlds shape the etiquettes (modes of comportment) that in turn shape our methods of understanding of the world. Our belief systems, or world views, arise in this context. It is this suggestion that I wish to explore in greater depth here.

It might be suggested that what I am getting at here is simply the widely accepted view that our reports of so-called facts are already, if implicitly, laden with theory, that cultural values are implicit in theory, and that these values are carried by the stories a culture tells about itself. Yes, this is *part* of it; but there is more to it, and there is more at stake.

N. Scott Momaday, in justly famous words, says: "It seems to me that in a sense we are all made of words; that our most essential being consists in language. It is the element in which we think and dream and act, in which we live our daily lives. There is no way in which we can exist apart from the morality of a verbal dimension."[14] Reading through Momaday's "The Man Made of Words," I came to realize that he was speaking not of sets of *beliefs* by which people constitute themselves, but more fundamentally of performance, enactment, the bringing into being of one's identity by means of action and practice, primary *verbal*. It is the difference between, say, the sacred as *object* of knowledge or belief (and, *derivatively*, of acts of faith and adoration) and sacramental practice—a matter of comportment, which in some sense brings into being a world, a ceremonial world, around it.

Ceremonial worlds are not fantasy worlds. We do, of course, experience the world. Experience is taken up into the ceremonial worlds. It is part of the self-correcting feedback loop that makes it possible for the day-to-day activities of food gathering, child rearing, shelter building, and so on to take place, to succeed, not only on the terms set by the world, but within the context of a richly textured ceremonial world. In such a world, as Paul Shepard has observed, "everyday life [is] inextricable from spiritual significance and encounter."[15]

The ceremonial world is the *center*. The kind of overall coherence for which ceremonial worlds strive is a mosaic of language (in the broadest sense) that serves many purposes at once. In the life of a community it must articulate a sense of those processes that bind the community together and to the land; and it must do this in a language that functions effectively to call forth appropriate responses. Above all, in such a world "natural things are not only themselves but a speaking."[16]

Sacramental *practice* is the key—not the *sacred*. What is important is ceremonial practice; it is that which defines the world in which we live and work. Etiquette is fundamental; it is in large part constitutive of one's mode of understanding the world. A "world view" is a kind of residue from one's ethical practice and the modes of attaining knowledge associated with that practice. This residue is highly prized, and receives intense scrutiny in academic contexts, but it is etiquette that is fundamental; etiquette is the fundamental dimension of our relationship to, and understanding of, the world. World views are at best pictures, metaphors, of ethical practice.

Storied Residence

Indigenous people not only acknowledge but often *celebrate* the differences that exist among the various Indigenous peoples in a truly remarkable way, one that has inclined me to prefer the term "ceremonial worlds" or "songs of the world" to "world views," which suggests the idea of a set of *beliefs* about how the world actually *is*. Truth figures in Indigenous worlds rather more obliquely than is suggested by "world view." As I have suggested, modes of understanding and comportment, etiquette, are intimately intertwined. This is perhaps summed up—and the ethical dimension given due emphasis—in Louise Profeit-LeBlanc's explication of the Northern Tutchone term *tli án oh* (klee-ah-no), often glossed as "what they say, it's true," as meaning "correctly true," "responsibly true" (a "responsible truth"), "true to what you believe in," "what is good for you and the community," and "rings true for everybody's well-being."[17]

Listen to Carol Geddes's account of the difference between approaching environmental ethics through theory (beliefs about the world) and through story (ceremonial worlds, songs of the world):

> I would like to tell you a small story about a very great lady in the Yukon. Her name is Mrs. Annie Ned. This illustrates, in a way, what bothers me about thinking about environmental ethics in the way we do today. Mrs. Annie Ned . . . was taken to a scientific conference in Kluane National Park. . . . Well, Mrs. Ned listened to all of the scientists giving their ideas about physical events in the park: what sort of things happened in the park, the geography of the park, and various other subjects. Mrs. Ned just very quietly listened to this all day. Then as they were leaving that evening . . . Julie [Cruikshank] said to Mrs. Ned, "How did you like the conference?"
>
> Mrs. Ned said to her, "They tell different stories than we do."
>
> This is very, very important, in fact it is profoundly important that we hear that. That is what they are, different stories. . . .
>
> [T]his is . . . a source of confusion for me: that I would be able to understand environmental ethics within the context of narrative as the way First Nations people were taught about the environment. We would never have a subject called environmental ethics; it is simply part of the story. When you are a child you first hear the animal mother story, about how animal mother gave the animals to the world, and how people have to consider this as a gift from the animal mother; and if we do not take care of the animals, then the animal mother will start to

take the animals back. We see that happening now. That is the context with which we understand environmental ethics, within that narrative, within the storytelling.

On the other hand there is all the scientific knowledge that we also learned in school, the different stories as Mrs. Ned said, the new paradigm. Too many people say, well let's take lessons from First Nations people, let us find out some of their rules, and let us try and adopt some of those rules. Let us try to look at it the same way that First Nations people do. But it is not something that you can understand through rules. It has got to be through the kind of consciousness that growing up understanding the narratives can bring to you. That is where it is very, very difficult, because people have become so far removed from understanding these kinds of things in a narrative kind of way.[18]

Theory is not inconsistent with storied understanding of self and world, but we must be careful to embed theory within our storied understanding of ourselves; we must make theory part of the story; give storied residence to our theories. And we must give *primacy* to story since stories provide a more nuanced, "ecological" understanding of our place in the world. They are the real homes of so-called thick moral concepts, the co-arising of fact and value. And stories *exemplify* what theory cannot, namely (as my colleague Lee Hester says), that in addition to the "true" statements that can be made about things, things have their *own* truth. To say that everything is its *own* truth is not so much a theoretical claim about the world as an expression of the thought that unless we extend a very basic courtesy to things in our attempts to understand them— or, as many indigenous people are inclined to put it (in English), unless we "respect" absolutely everything—we cannot arrive at an understanding of them *or* ourselves that makes *sense*, that makes sense of our lives, our cultures, our relationship to all that is. We cannot otherwise know what is *tli án oh*, what is "correctly true."

An elder telling Papago origin stories at a meeting about educational programs for Indigenous people constructs a world in which discussion can meaningfully proceed.[19] Yukeoma begins his argument for why Hopi children shouldn't attend white schools by "speaking within the framework of the Hopi origin saga and its prophecies."[20]

In contrast, although most of us in the West know that theories are deeply (though implicitly) shaped by personal and cultural values and that these values are deeply (though again implicitly) shaped by stories—that is, they are carried and propagated by the stories that define us as individuals and define the cultures within which we live and come to understand ourselves—this understanding is not often (or often enough) reflected in practice; nor is it reflected in our meta-level analyses of practice. This chapter has been no exception. I did not begin with an origin myth; I did not first invoke a world within which discussion might meaningfully proceed. I have pointed to the practice of Indigenous people who *do* begin in this way, but these illustrations were not situated within stories that defined the world *I* inhabit; I have been speaking as if from *nowhere* or *everywhere*, thus illustrating the very theoretical tendency I have been at pains to criticize in this chapter. All too often the implicit assumption underlying our discussions in environmental ethics is that we can profitably discuss these matters without defining and locating the ceremonial worlds (the stories) within which our discussions proceed. We speak as though *from* no world at all; and we presumptuously speak *for* all worlds.

As environmental ethicists, we might begin to explicitly discover and acknowledge the stories within which we think about environmental ethics. Only then can we begin to tell (and live within) better stories, stories that lead us away from the human centeredness that defines Western culture. Only then can we begin to live "in the presence of the more-than-human . . . to awake and go to sleep with it, to take its rhythms and cycles for the rhythms and cycles of [our own lives], until the two finally merge into one stream," as Anthony Weston has so eloquently put it.[21]

In this matter of story telling, most of us need edification—that is, moral and (dare I say it?) spiritual instruction. This we can find in the cultural worlds of Indigenous peoples. The stories that define Indigenous worlds intertwine the sacred, the natural, and the personal; they mediate the paradoxes of existence. The real world is a ceremonial world in which animals are kin, in which food and knowledge come as gifts from powers beyond us, and the human role in the scheme of things is at once ceremonial and practical. If we were to refract our thought concerning environmental ethics through the lens of such a world, we would come to understand that environmental ethics is really a spiritual concern. That is to say, environmental ethics would be understood as concerned with our relationship to a more-than-human world, and as enjoining a mindfulness that could be called "walking in a sacred manner."

As we have seen, Momaday says, "It seems to me that in a certain sense we are all made of words; that our most essential being consists in language. It is the element in which we think and dream and act, in which we live our daily lives. There is no way in which we can exist apart from the morality of a verbal dimension."[22]

We are all made of words; and we have our existence within stories, the sacred and profane stories that constitute our cultural and personal identities. For most of us, these stories (these actions that create and sustain our worlds) touch on our biological and ecological existence only incidentally. But our existence is deeply ecological, and our cultural identities should reflect this, as do those of Indigenous peoples. "The mythtellers speak of the powers *in relation* to each other, and with an eye to the whole ecology, not separable functions of it,"[23] as in Skaay's coda to a long and complex narrative poem:

> This place was round,
> and grass surrounded it, they say.
> He travelled around in it, Sapsucker did.
> Every one of his feathers was missing.
> Up above, there was a big spruce sloughing off its skin.
>
> He whacked it with his beak.
> And as he drummed his beak against it,
> something said,
> "Your father's father asks you in."
> He looked for what had spoken.

Nothing stood out.
After something said the same thing again,
he looked inside the hollow of the tree.
Someone shrunken and sunken, white as a gull, sat at the
    back.
Then he went in.

The elder reached into a small box.
After he had pulled five boxes out from inside one another,
he showed him his wing feathers.
Oooooooooooh my!
Then he gave him tailfeathers too.

Then he shaped him with his hands.
He colored the upper part of him red,
and then he said to him,
"Now, my grandson, you should go.
This is why you have been with me."

Then he went back out,
and then he flew,
and then he did the same thing as before.
He clutched the tree,
and then he struck it with his beak.

And so it ends.[24]

Sean Kane comments, "each time a Red-headed Sapsucker drills its
parallel rows of identical holes in a spruce tree, then comes back
to feed on the insects drawn to the sap on that tree, the bird takes
with that food the spiritual energy of the Old Man who is his kin.
'Your father's father asks you in.' And the Old Man is reborn in
the beauty of the Sapsucker for, according to the logic of Haida
kinship, the grandfather is reincarnated in the grandson. Life is
served . . . knowledge of pattern is the beginning of every practical wisdom."[25]

Jeannette Armstrong adds an ecological dimension to Momaday's thought that we are all made of words: "The Okanagan word for 'our place on the land' and 'our language' is the same. The Okanagan language is thought of as the 'language of the land.' This means that the land has taught us our language. The way we survive is to speak the language that the land offered us as its teachings. . . . We also refer to the land and our bodies with the same root syllable. . . . We are our land/place."[26]

We find a Western echo of this connection between language, land, and self in Conrad Aiken's words: "The landscape and the language are the same. / And we ourselves are language and are land."[27]

The landscapes that shape most of our identities, however, are human landscapes, landscapes of human culture and humanly transformed nature—broken landscapes that mirror our own brokenness.

This has not always been so and is even now not so for perhaps most Indigenous peoples. The *deepest* sources of personal and cultural identity are the ecological and geological landscapes that shape and sustain us. This, and our present loss, are given voice in what are surely Momaday's most memorable words:

> East of my grandmother's house the sun rises out of the plain. Once in his life a man ought to concentrate his mind upon the remembered earth, I believe. He ought to give himself up to a particular landscape in his experience, to look at it from as many angles as he can, to wonder about it, to dwell upon it. He ought to imagine that he touches it with his hands at every season and listens to the sounds that are made upon it. He ought to imagine the creatures there and all the faintest motions of the wind. He ought to recollect the glare of noon and all the colors of the dawn and dusk.[28]

These unbroken landscapes are characterized by their integrity. As Barry Lopez has put it: The "landscape is organized according to principles or laws or tendencies beyond human control. It is understood to contain an integrity that is beyond human analysis and unimpeachable."[29]

It is this integrity, beyond human analysis and unimpeachable, that marks the land as sacred for most Indigenous peoples. The "sacred" (for example, the Lakota *wakan tanka*, "great mysterious") is the more-than-human quality of the world, not a being transcendent to the world. Asking about the meaning and origin of the term *wakan tanka* a Lakota asked older Lakota about its meaning and origin. He received this story as his answer:

> Way back many years ago, two men went walking. It was on the prairies. As they walked, they decided, "Let's go up the hill way towards the west; let's see what's over the hill."
> So they walked and they came to the top of this hill and they looked west and it was the same. Same thing as they saw before; there was nothing. They just kept going like that, all day and it was the same. They came to a big hill and there was another big hill further back. Finally they stopped and they said, "You know, this is Wakan Tanka."[30]

In this sense of "sacred"—in which the sacred is a dimension of the natural landscape itself—evolutionary biology and ecosystem ecology may provide us with sacred myths of origin, for they portray the world as more-than-human and evoke a deep and inclusive sense of kinship with it. Nature's complexity, its generosity, and its communicative ability make it possible for us to experience the deep unity of the personal, the sacred, and the natural.

But all this is quite abstract and doesn't speak to us of the particularities of our homes, our places on earth. Within many Indigenous cultures, myths of origin and other stories are ceremonial creations and renewals of worlds that tie cultural identity, even survival itself, to specific landscapes. Simon Ortiz, in three poems from *A Good Journey*, writes:

SURVIVAL THIS WAY

Survival, I know how this way.
This way, I know.
It rains.
Mountains and canyons and plants
grow.

We traveled this way,
gauged our distance by stories
and loved our children.
We taught them
to love their births.
We told ourselves over and over
again.
"We shall survive this way."

SPEAKING

I take him outside
under the trees,
have him stand on the ground.
We listen to the crickets,
cicadas, million years old sound.
Ants come by us.
I tell them,
"This is he, my son.
This boy is looking at you.
I am speaking for him."

The crickets, cicadas,
the ants, the millions of years
are watching us,
hearing us.
My son murmurs infant words,
speaking, small laughter
bubbles from him.
Tree leaves tremble.
They listen to this boy
speaking for me.

CANYON DE CHELLY

Lie on your back on stone,
the stone carved to fit

the shape of yourself.
Who made it like this,
knowing that I would be along
in a million years and look
at the sky being blue forever?

My son is near me. He sits
and turns on his butt
and crawls over to the stones,
picks one up and holds it,
and then puts it into his mouth.
The taste of stone.
What is it but stone,
the earth in your mouth.
You, son, are tasting forever.

We walk to the edge of cliff
and look down into the canyon.
On this side, we cannot see
the bottom cliff edge but looking
further out, we see fields,
sand furrows, cottonwoods.
In winter, they are softly gray.
The cliffs' shadows are distant,
hundreds of feet below;
we cannot see our own shadows.
The wind moves softly into us.
My son laughs with the wind;
he gasps and laughs.

We find gray root, old wood,
so old, with curious twists
in it, curving back into curves,
juniper, piñon, or something
with hard, red berries in spring.
You taste them, and they are sweet
and bitter, the berries a delicacy

for bluejays. The plant rooted
fragiley into a sandy place
by a canyon wall, the sun bathing
shiny, pointed leaves.

My son touches the root carefully,
aware of its ancient quality.
He lays his soft, small fingers on it
and looks at me for information.
I tell him: wood, an old root,
and around it, the earth, ourselves.[31]

Other stories, more familiar to Euro-Americans, also define us
in relationship to the land. One such tale is Aldo Leopold's "Marsh-
land Elegy," in his *A Sand County Almanac.*

A dawn wind stirs on the great marsh. With almost imper-
ceptible slowness it rolls a bank of fog across the wide morass.
Like the white ghost of a glacier the mists advance, riding over
phalanxes of tamarack, sliding across bogmeadows heavy with
dew. A single silence hangs from horizon to horizon. . . .
   A sense of time lies thick and heavy on such a place. Yearly
since the ice age it has awakened each spring to the clangor of
cranes. The peat layers that comprise the bog are laid down in
the basin of an ancient lake. The cranes stand, as it were, upon
the sodden pages of their own history. These peats are the com-
pressed remains of the mosses that clogged the pools, of the
tamaracks that spread over the moss, of the cranes that bugled
over the tamaracks since the retreat of the ice sheet. An endless
caravan of generations has built of its own bones this bridge
into the future, this habitat where the oncoming host again may
live and breed and die.
   To what end? Out on the bog a crane, gulping some luck-
less frog, springs his ungainly hulk into the air and flails the
morning sun with mighty wings. The tamaracks re-echo with
his bugled certitude. He seems to know. . . .
   This much, though, can be said: our appreciation of the crane
grows with the slow unraveling of earthly history. His tribe, we
now know, stems out of the remote Eocene. The other members

of the fauna in which he originated are long since entombed within the hills. When we hear his call we hear no mere bird. He is the symbol of our untamable past, of that incredible sweep of millennia which underlies and conditions the daily affairs of birds and men.

And so they live and have their being—these cranes—not in the constricted present, but in the wider reaches of evolutionary time. Their annual return is the ticking of the geologic clock. Upon the place of their return they confer a peculiar distinction. Amid the endless mediocrity of the commonplace, a crane marsh holds a paleontological patent of nobility, won in the march of eons. . . . The sadness discernible in some marshes arises, perhaps, from their once having harbored cranes. Now they stand humbled, adrift in history.[32]

This story helps define me and my neighbors in southeastern Wisconsin, in relationship to the geologic and ecosystemic legacy of the last Wisconsin Ice, as a prairie/wetland people. The elegy also haunts us; it is a story of loss. This fits our cultural temper. It is a fair question whether our religions of loss and redemption are in some way tied to the mutual estrangement of the natural, the personal, and the sacred in Western culture.

The search for roots can take other shapes than that of a search for redemption in the mode of a search for the "Truth" of one's origin and identity. The following example comes from my colleague, Lee Hester. The Choctaw people migrated long ago to Mississippi carrying the bones of their ancestors with them. When they reached Mississippi they are said to have built the mound of *Nanih Waiyah* to house these bones. Yet *Nanih Waiyah* is also said to be the great "Productive Mound" from which all people emerged. From the point of view of the "One (literal) Truth" this seems contradictory; the *new* burial mound *couldn't* be the Choctaw place of *origin*, emergence. From the point of view of Choctaw *practice*, however, a different meaning of emergence and origins arises.

When forcibly removed to Oklahoma, the first Choctaw capitol in Oklahoma was called *Nanih Waiyah*; and even today there is a lake *Nanih Waiyah* near the Choctaw Nation of Oklahoma council building at Tuskahoma. Though they were forced to leave their

ancestors behind, and though many of their loved ones died on the
Trail of Tears, the Choctaw people of Oklahoma are rooted in their
new land. Choctaw practice has the consequence that as a people
the Choctaw are always at home on this earth, never detached from
tradition and tribal history—these are always present in the tan-
gible form of the emergence mound. Practice and the social meaning embedded in that practice are central.

I would like to think that, should we Euro-Americans come to
experience—genuinely experience—the ecological integrity of the
land in a way that might define us culturally and personally, that
redemption, in one of its aspects at least, would be closer to hand
than it has been in the redemptive religions of the West and the
Orient. Alienation would not be experienced as a cosmic sunder-
ing of God and his people, but rather, Barry Lopez has put it, as
a "persistent principle of disarray" checked by ceremonies of renewal that make the individual "a reflection of the myriad enduring relationships of the landscape."[33]

As an example of such ceremonial renewal and regeneration, I
speak some of the prairie restoration project my family is involved
with at the Waterville Prairie, University of Wisconsin Biological
Field Station.

In 1978 Fran and I returned to Wisconsin from a year on a salt
marsh on Long Island Sound to two important and enduring
changes in our lives: the adoption of our son, Carlos, and our in-
volvement in the Waterville Prairie restoration project.

Twice in the Autumn—late September and late October—we
gather with other friends of the field station to gather seeds from
remnants of the tallgrass prairies that covered significant portions
of southeastern Wisconsin before the plow. In April, we burn the
prairie, something formerly accomplished by natural wildfires and
fires deliberately set by Indigenous peoples to prevent the spread
of forests, thereby keeping the land open for game. Our prairie
burns accomplish several things: they prevent the spread of forests
onto the prairie, as I have said; they set back the growth of cool-
weather European grasses and forbs, providing the native prairie
species a place in the sun; they release nutrients into the soil; and
they warm the soil, encouraging the warm-weather native grasses

and forbs. In May, we plant new sections of prairie, setting out seedlings and raking seeds into the rich prairie soil.

The activities of these four days have become important ceremonials in our lives. The scale of this restoration project is ceremonial, it is culturally and personally regenerative. These are world-renewal ceremonies—in some measure they define and shape our community, our world—and they are healing ceremonies, realigning us with land that in some measure defines us personally. Our relationship to this place is a defining relationship. It is a powerful place of intersection for Fran's work as artist, my work as teacher and our personal, family, and community lives. It is a place that orients us morally and spiritually in this world. The prairie ceremonials are enactments of fundamental principles of moral attentiveness to all that exists. Our biological myths of origin (evolutionary biology and ecosystem ecology) give broad philosophical direction and content to these acts of mindfulness.

The following remarks by Gary Snyder, whose chosen place on earth is on the west slope of the Sierra Nevada range in northern California, strike the right tone, I think. They articulate nicely the sense of the word sacred as I have been using it.

These foothill ridges are not striking in any special way, no postcard scenery, but . . . the fact that my neighbors and I and all of our children have learned so much by taking our place in these Sierra foothills—logged-over land now come back, burned-over land recovering, considered worthless for decades—begins to make this land a teacher to us. It is the place on earth we work with, struggle with, and where we stick out the summers and winters. It has shown us a little of its beauty.

And sacred? One could indulge in a bit of woo-woo and say, yes, there are newly discovered sacred places in our reinhabited landscape. I know my children (like kids everywhere) have some secret spots in the woods. There is a local hill where many people walk for the view, the broad night sky, moon-viewing, and to blow a conch at dawn on Bodhi Day. There are miles of mined-over gravels where we have held ceremonies to apologize for the stripping of trees and soil and to help speed the plant-succession recovery. There are some deep groves where people got married.

Even this much connection with the place is enough to inspire the local community to hold on: renewed gold mining and stepped-up logging press in on us. People volunteer to be on committees to study the mining proposals, critique the environmental impact reports, challenge the sloppy assumptions of the corporations, and stand up to certain county officials who would sell out the inhabitants and hand over the whole area to any glamorous project. It is hard, unpaid, frustrating work for people who already have to work full time to support their families. . . . More than the logic of self-interest inspires this: a true and selfless love of the land is the source of the undaunted spirit of my neighbors.

There's no rush about calling things sacred. I think we should be patient, and give the land a lot of time to tell us or the people of the future. The cry of a Flicker, the funny urgent chatter of a Grey Squirrel, the acorn whack on a barn roof—are signs enough.[34]

Here I think of images nested in Fran's baskets and necklaces: Autumn sky and Big Bluestem; Summer Cooper's Hawks nested in the woods; Canada Geese flying in low; Autumn seed gathering—these, too, are signs enough.

More than one tradition relates that our human purpose in this world is to tell stories. The animal and plant people decided that they would provide what we humans need so that we may tell the stories needed to create and continually renew this sacred world. Our stories of the land, the more-than-human dimension of our being, should be true sacraments—outward and visible signs of inward and spiritual grace. The notion of the sacred gives way to the notion of sacramental practice, walking in a sacred manner.

Notes

1. Henry S. Sharp, *The Transformation of Bigfoot: Maleness, Power, and Belief among the Chipewyan* (Washington D.C.: Smithsonian Institution Press, 1988), pp. 144–45.

2. Evelyn Fox Keller, *Reflections on Gender and Science* (New Haven and London: Yale University Press, 1985), p. 134.

3. Donna Haraway, "Situated Knowledges: The Science Question in Feminism and the Privilege of Partial Perspective," *Feminist Studies* 14 (Fall 1988), pp. 593–94.

4. Frederick Nietzsche, *The Gay Science*, §114.

5. On "universal consideration" see Thomas H. Birch, "Moral Considerability and Universal Consideration," *Environmental Ethics* 15 (Winter 1993).

6. Carol Geddes, panel discussion by Yukon First Nations people on the topic of "What is a good way to teach children and young adults to respect the land?" transcript in Bob Jickling, editor, *Environment, Ethics, and Education: A Colloquium* (Whitehorse: Yukon College, 1996), p. 46.

7. Quoted in Gerald Vizenor, "Trickster Discourse: Comic & Tragic Themes in Native American Literature," in Mark Lindquist and Martin Zanger, eds., *"Buried Roots and Indestructible Seeds": The Survival of American Indian Life in Story, History, and Spirit* (Madison: The Wisconsin Humanities Council, 1993), p. 39.

8. Lee Irwin, *The Dream Seekers: Native American Visionary Traditions of the Great Plains*, foreword by Vine Deloria, Jr. (Norman: University of Oklahoma Press, 1994), p. 224.

9. Walter J. Ong, S.J., "World as View and World as Event," *American Anthropologist* 71 (1969), pp. 637, 638, 636.

10. Sam Gill, "Holy Book in Nonliterate Traditions: Toward the Reinvention of Religion," in Sam Gill, *Native American Religious Action: A Performance Approach to Religion* (Columbia: University of South Carolina Press, 1987), pp. 139–40.

11. Sam Gill, "One, Two, Three: The Interpretation of Religious Action," in Gill, *Native American Religious Action*, pp. 162–63, 151.

12. Gill, "One, Two, Three," p. 162.

13. Diamond Jenness, "The Carrier Indians of the Bulkley River," *Bureau of American Ethnology Bulletin* No. 133 (Washington, 1943), p. 540. As quoted in Gill, "Holy Book," p. 131.

14. N. Scott Momaday, "The Man Made of Words," in Sam Gill, *Native American Traditions: Sources and Interpretations* (Belmont, CA: Wadsworth Publishing Company, 1983), p. 44.

15. Paul Shepard, *Nature and Madness* (San Francisco: Sierra Club Books, 1982), p. 6.

16. Shepard, *Nature and Madness*, p. 9.

17. In conversation at the Colloquium on Environment, Ethics, and Education, Yukon College, Whitehorse, Yukon, July 14–16, 1995.

18. Geddes, panel discussion, in Jickling, *Environment, Ethics, and Education*, pp. 32–33.

19. Sam Gill, "The Trees Stood Deep Rooted," in *Native American Religious Action*, p. 17.

20. Nabokov, "Present Memories, Past History," pp. 148–49.

21. Anthony Weston, *Back to Earth: Tomorrow's Environmentalism* (Philadelphia: Temple University Press, 1994), p. 143.

22. Momaday, "The Man Made of Words," p. 49.

23. Sean Kane, *Wisdom of the Mythtellers* (Peterborough, Ontario: Broadview Press, 1994), p. 36.

24. This three-part narrative poem is itself just the second of five parts of a larger narrative cycle, *The Qquuna Cycle*, that took Skaay of the Qquuna Qiighawaai most of the month of October, 1900 to dictate. Translated by Robert Bringhurst in chapter 5 of his *A Story as Sharp as a Knife: An Introduction to Classical Haida Literature* (Vancouver/Toronto: Douglas & McIntyre, 1998). Bringhurst's earlier translation of the coda, along with the original Haida text, appears in Kane, *Wisdom of the Mythtellers*, pp. 28–31.

25. Kane, *Wisdom of the Mythtellers*, pp. 36–37.

26. Jeannette Armstrong (Okanagan), "Keepers of the Earth," in *Ecopsychology: Restoring the Earth, Healing the Mind*, ed. Theodore Roszak et al. (San Francisco: Sierra Club Books, 1995), p. 323.

27. Conrad Aiken, quoted in Edith Cobb, *The Ecology of Imagination in Childhood* (New York: Columbia University Press, 1977), p. 67.

28. N. Scott Momaday, *The Way to Rainy Mountain* (New York: Ballantine Books, 1970), p. 113. I retain the first sentence of the section, usually dropped when quoted.

29. Barry Lopez, "Landscape and Narrative," in Lopez, *Crossing Open Ground* (New York: Random House, 1988), p. 66.

30. Elaine Jahner, "The Spiritual Landscape," in D. M. Dooling and Paul Jordan-Smith, editors, *I Become Part of It: Sacred Dimensions in Native American Life* (San Francisco: HarperCollins, 1992), p. 193.

31. Simon Ortiz, *A Good Journey*, collected in *Woven Stone* (Tucson: The University of Arizona Press, 1992), pp. 167–8, 190, 201–2. "Survival This Way" is a section of a longer poem, "A San Diego Poem: January–February 1973."

32. Aldo Leopold, *A Sand County Almanac* (New York: Ballantine Books, 1970), pp. 101–3.

33. Lopez, "Landscape and Narrative," pp. 67–68.

34. Gary Snyder, *The Practice of the Wild* (San Francisco: North Point Press, 1990), pp. 95–96.

Epilogue
Going On

Each in its different way, the essays in this book aim to open up new ways of thinking of ourselves and of more-than-human worlds, and to open up new possibilities for living within and with them. The aim of this epilogue is to offer some ways of going on from here, both to deepen and extend the questions raised so far and to move outward from classroom and library—to integrate the new questions and new ways of thinking with practical action. The field is *vast*; we can barely touch down at some of the key places.

I hope readers will at least find here a little concrete guidance—
some directions to go next, some sources and resources—and,
above and beyond all the details, a sense also of just how many and
how wild the possibilities really are.

Coming to Our Senses

Abram's essay and my own both insist that our deepest need is to
recover some sense—quite literally some *sense*—of the larger world.
Returning to nature is not merely, and not primarily, an intellectual step. At some point we must lift our heads from our books
and lift ourselves from our desks, and sing with the birds and sleep
on the open ground.

So learn the birds, the stars, the trees, the shells, the bugs. Field
guides on all of these subjects, and many more, can be found in
most bookstores. There are junior versions for the kids. Look for
local backpacking or wilderness camping groups. Join the Sierra
Club (730 Polk Street, San Francisco, CA 94109), which has local
affiliates in most areas that not only undertake the usual lobbying
and organizational work but also create outdoor opportunities, like
trail repair trips and campouts. Look for local outing groups.

Pay attention to the animals. I ask my environmental philosophy students to pick an animal (or plant, or place) with which they
identify, and learn enough to tell the rest of us about it, bringing
in tokens too if they can. In the end they speak as that animal (or
. . .) in a closing conclave that attempts on a small scale what John
Seed calls a "Council of All Beings." See John Seed, Joanna Macy,
Pat Fleming, and Arne Naess, *Thinking Like a Mountain: Towards
a Council of All Beings* (New Society Publishers, 1988).

Pay attention to the great cycles of light and dark, to the rhythm
of the seasons. Our contemporary holidays are directly linked to
the solstices and equinoxes and the "cross-quarter days" in between, all well known to the Celts for example, from whom we
even have the names "Easter" (Spring Equinox: the time of resur-
rection, as Earth returns to life) and "Yule" (Winter Solstice: when
the sun begins to come north again, hence the rebirth of the year

and the sun itself, the promise of salvation in the midst of the cold and dark winter). These are portentous times of year, which we can recognize and celebrate regardless of whether we are Christian or Jew or not religious at all. For detailed examples of seasonal rituals, see Donna Henes, *Celestially Auspicious Occasions: Seasons, Cycles, and Celebrations* (Berkeley Publishing Group/Perigee), and Luisah Teish, *Carnival of the Spirit: Seasonal Celebrations and Rites of Passage* (Harper).

Imagine the other holidays we could have too. Already at New Year's many people all across the country traipse out before dawn to count birds for the Audubon Society. To join, watch your local paper, or contact the National Audubon Society (700 Broadway, NY, NY 10003). Why don't we make Bird Count Day a national holiday? Imagine weeks of preparation by eager schoolchildren learning to identify birds. Imagine "Star Nights" on which all lights everywhere are turned out: these could be timed to coincide with meteor showers, eclipses, occlusions. The poet Antler recalls Ralph Waldo Emerson's epiphany—"If the stars came out only one night in a thousand years, how people would believe and adore, and preserve from generation to generation, remembrance of the miracle they'd been shown"—and imagines the scene:

> Whole populations thronging to darkened
>   baseball stadiums and skyscrapertops
>   to sit holding hands en masse
>   and look up at the billion-year spree
>   of the realm of the nebulae!

Much "nature writing" has the experience of nature as its main focus. For a survey, see *The Norton Book of Nature Writing,* edited by John Elder (W. W. Norton, 1980). For an invitation specifically to nature poetry, see Robert Bly, ed., *News of the Universe* (San Francisco: Sierra Club Books, 1980). A striking recent anthology of nature writing is *The Sacred Place* (University of Utah Press, 1996), edited by W. Scott Olsen and Scott Cairns, bringing together writers and poets who convey the sense that in some of our experience of nature there is "something 'bigger' going on, something more significant at stake." "The works collected here share a common reverence for *the world itself*, and—perhaps best of all—they share a common understanding that no one of them comprehends fully what that means" (p. xii).

For one absolutely vital step, very practical and close to home—turn off the TV. In fact, get rid of it. Make it into a planter. My essay "Is It Too Late?" launches a first challenge to the current role of television in our lives, but there is much, much more. Writer Bill McKibben watched TV for a week and then camped for a week by himself in the Adirondacks. For his account of the difference and the difference it makes, see *The Age of Missing Information* (Random House, 1992). To connect plug-pulling to larger environmental and cultural themes, see Jerry Mander's classic *Four Arguments for the Elimination of Television* (NY: Morrow, 1978) and, painted on a larger canvas, his *In the Absence of the Sacred: The Failure of Technology and the Survival of the Indian Nations* (Sierra Club, 1991). For moral support, contact TV-Free America (1611 Connecticut Avenue NW, Suite 3A, Washington, DC 20009).

Environmental Education

Disconnection begins young. In a recent survey of U.S. fifth and sixth graders, 53 percent of the children listed the media as their primary teacher about nature, 31 percent cited school, and only 9 percent cited learning at home and actual experience outside. Mander, just cited, is concerned for this. See also Gary Nabhan and Steven Trimble, *The Geography of Childhood: Why Children Need Wild Places* (Beacon Press, 1994). Children *need* wild places, they argue, streams to wander and mud and the unexpected animal; and this is something that adults now need to attend to and offer. See also Roger Hart, *Children's Participation: The Theory and Practice of Involving Young Citizens in Community Development and Environmental Care* (UNICEF: Earthscan Publications, 1997).

On environmental education at the primary and secondary levels, see Barbara Robinson and Evelyn Wolfson, *Environmental Education: A Manual for Elementary Educators* (New York: Teacher's

College Press, 1982). Look for local resources. Poking around my local State University library, I found a whole array of elementary teaching support materials, exploring everything from sea turtles and beavers to biting plants to geological time. Much of this material is tied in to specific state parks and developed in concert with them. On environmental education at the college level, see David Orr, *Earth in Mind* (Island Press, 1994) and, for a practical guide, Jonathan Collett and Stephen Karakshian, *Greening the College Curriculum: A Guide to Environmental Teaching in the Liberal Arts* (Island Press, 1996). A good contact for environmental education on all levels is the North American Association for Environmental Education (1255 23rd St NW Suite 400, Washington, DC 20037).

Delve deeper into your own place. Half the place-names on Long Island, where I used to teach, are Native American: Setauket, Montauk, Massapequa, and Manhattan itself. An entire history is there to be recovered. The teams from my southern Wisconsin high school (in a town called Spring Green, a school district called River Valley—listen to the names!) were the Blackhawks: the name turns out to go back to the Sauk war chief Black Hawk— not that we were ever told this in high school—whose rebellion led him through those parts in 1837. Spring Green, just downriver from Sauk City, the fertile riverbank that was the Sauks' City. The Sauk ended up in Kansas; the first white settlers in Wisconsin came to be called "badgers"—subsequently the University of Wisconsin totem animal—after their style of winter squatting on Sauk land, burrowed into the ground imitating the real badgers. The Sauk lived in houses. All of this, though, is a long story, or rather many stories. *Every* place has such stories. Every place has such a history. What about yours?

Near my home now in North Carolina there is a festival every spring in which musicians, ecologists, and storytellers paddle down the Haw River, stopping at every town to teach the ecology and history of the river to the children—it is one of the highlights of the local 4th grades—and to promote general environmental awareness. Check alternative bulletin boards for similar opportunities near you. In the library, look for local natural histories.

Ecology and the Environmental Sciences

Increasingly American colleges and universities are offering inter-disciplinary courses of study in environmental science and environmental studies. Environmental *science* programs tend to offer a range of science courses (chemistry, biology) tied together from an ecological point of view. Environmental *studies* programs try to integrate a wider range of disciplines, including political science, economics, religion, and philosophy. Both types of program are geared toward the increasing numbers of jobs related to environmental concerns: assessing environmental impacts, planning on all levels of government, policy-setting and politics. See Bill Sharp, et al., *The Complete Guide to Environmental Careers* (Island Press: regular re-issues).

For an especially accessible introduction to ecological science, see G. Tyler Miller, *Environmental Science* (Wadsworth, many editions). For an intriguing intellectual history of ecological science, see Donald Worster, *Nature's Economy* (Cambridge University Press, second edition 1994). Worster points out that there has always been a side of ecology that has been activist, not content with theory's ivory tower. Miller's book itself can be read as a sustained and well-buttressed piece of advocacy. For direct applications of conservation biology to ecological policy, one good example is the extensive booklist offered by Island Press (Box 7, Covelo, CA 95428).

The final chapters of the second edition of Worster's *Nature's Economy* detail some striking new developments in ecology. Ecologists are now even challenging some of the notions of ecology that have just begun to enter the public consciousness, such as the "balance of nature." See also Worster's "The Ecology of Order and Chaos," in his *The Wealth of Nature* (Oxford University Press, 1993). Linking some of this rethinking to developments in postmodernism is Michael Soule and Gary Lease, *Reinventing Nature? Responses to Postmodern Deconstruction* (Island Press, 1995).

The scientific literature in ecology, biology, and the other life sciences is, of course, vast. On biodiversity, standouts are E. O. Wilson's *The Diversity of Life* (Harvard University Press, 1992) and Reed Noss's *Saving Nature's Legacy: Protecting and Restoring Biodi-*

versity (Island Press, 1994). On the interlacing webs of life in movement that link the globe and all of life, start with the birds: see John Terborgh, *Where Have All the Birds Gone? On the Biology and Conservation of Birds that Migrate to the American Tropics* (Princeton University Press, 1990) and Jonathan Elphick, general editor, *The Atlas of Bird Migration* (Random House, 1995). Related is Stephen Buchmann and Gary Paul Nabhan, *The Forgotten Pollinators* (Island Press, 1996)—the reference is to bees, bats, butterflies, and a great many other animals, many threatened, many unappreciated. "More than any other natural process, plant-pollinator relationships offer vivid examples of the connections between endangered species and threatened habitats. . . ."

But these are only a few examples of literally thousands. Find a good library, pick an area of interest or importance to you and your community, and start there.

Beginning at the various boundaries of the scientific literature are a variety of other literatures, again enormous in many cases. Some integrate natural and cultural history, as in many works of Barry Lopez, for example his *Arctic Dreams: Imagination and Desire in a Northern Landscape* (Bantam, 1987). Lyall Watson's essays in his *Earthworks: Ideas on the Edge of Natural History* (London: Hodder and Stoughton, 1986), "each take an odd idea, something from the soft edges of science, and try to nourish it with natural history—to work it, somehow, into the fabric of Earth." One astonishing example is Watson's *Gifts of Unseen Things* (Colchester, VT: Destiny Books, 1991). A modern scientific and cultural travelogue is Mark Plotkin's *Tales of a Shaman's Apprentice* (Viking, 1993).

On the powers of animals, there is again a breathtaking range of reading: everything from the scientific (see Donald Griffin *Animal Minds* (University of Chicago Press, 1992)) to the lyrical and poetic (for example Tom Jay's marvelous essay "The Salmon of the Heart" (in Finn Wilcox and Jeremiah Gorsline, eds., *Working the Woods, Working the Sea* (Port Townsend, WA: Empty Bowl Press, 1986)). On dogs, see Elizabeth Marshall Thomas, *The Hidden Life of Dogs* (Houghton Mifflin, 1993). On monkeys, see Dorothy Cheney and Robert Seyfarth, *How Monkeys See the World* (University of Chicago Press, 1990). On cetaceans, still among the best is Joan McIntyre, ed., *Mind in the Waters* (Sierra Club, 1974); see also, on whales and a variety of other animals, Diane Ackerman's *The Moon by Whalelight* (Random House, 1991). Jim Nollman's *Animal Dreaming: The Art and Science of Interspecies Communication* (Bantam Books, 1987) is cited at length in my essay in this book. More works keep appearing. Jeffrey Moussaieff Masson, in *When Elephants Weep: The Emotional Lives of Animals* (NY: Delacorte Press, 1995), weaves together a wide range of research dealing with animal emotion—with play, fear, love, grief, shame, even artistic creation and worship—into one compelling account.

The State of the World

For general and up-to-date information on the global situation, see the Worldwatch papers, such as the *Vital Signs* annual series, published by the Worldwatch Institute (1776 Massachusetts Avenue NW, Washington, DC 20036); and the "Environment" Annual Edition of the Dushkin Publishing Company's *Annual Editions Series* (Dushkin Publishing Group, Inc., Sluice Dock, Guilford, CT 06437). On the World Wide Web, try the Environmental News Network at <http://www.enn.com>, and (a general resource) Envirolink Library at <http://www.envirolink.org/EnviroLink_Library/>.

There is some debate about just how bad things are. Some critics argue that environmentalism *radically* overstates the dangers. A representative anthology of the critics is Julian Simon, ed., *The State of Humanity* (Blackwell, 1995). See also Norman Myers, *Scarcity or Abundance? A Debate on the Environment* (Norton, 1994).

Historically speaking, though, no one denies that the human entry onto the scene, and the rise of industrialism in paticular, meant massive changes, usually depletions and extinctions, for the natural world as it existed earlier, as well as for native peoples. See for example Farley Mowat, *Sea of Slaughter* (Atlantic Monthly Press, 1984), an account of the spectacular destruction wrought by Europeans in the North Atlantic. Bullfrog Films (Oley, PA 19547) sells a stunning video, narrated by Mowat, under the same title. Mowat is Canada's leading nature writer; see also his *A Whale for the Killing* (Bantam, 1972), in which he struggles to find words for a feeling of profound kinship with a trapped whale, "alien flesh calling out to alien flesh." For a similar argument painted on a much broader canvas, see Clive Pointing, *A Green History of the World: The Environment and the Collapse of Great Civilizations* (St. Martin's Press, 1992). Alfred Crosby's *Ecological Imperialism: The Biological Expansion of Europe 900–1900* (Cambridge University Press, 1986) argues that "Europeans' displacement and replacement of the native peoples in the temperate zones was more a matter of biology than military conquest."

A great deal of popular environmentalism is in the key of loss and danger: one fine representative is Albert Gore's (that's *Vice President* Albert Gore's) *Earth in the Balance* (Houghton Mifflin, 1992). For one pessimistic view of the likely end result of human exploitation of nature, see Bill McKibben, *The End of Nature* (Random House, 1989). A very good reply to *The End of Nature* is, oddly enough, the very same Bill McKibben's *Hope: Human and Wild* (Little, Brown, 1995).

Sources of the Crisis

Another philosophical topic is *how all of this happened* in the first place: not just the shape of the crisis, hard as it may be to bring *that* into focus, but its ultimate sources and causes.

We tend to see the environmental crisis as a new event, but many scholars see it as much older, even ancient. Even the early Mediterranean world was aware of suffering ecological damage: see J. Donald Hughes, *Ecology in Ancient Civilizations* (University of New Mexico Press, 1975). Paul Shepard, in *Nature and Madness* (Sierra Club, 1982), traces our environmental alienation all the way back to the prehistoric shift from hunting and gathering to settled agriculture. Agriculture, vulnerable and simplified, defines everything as for or against, and moreover nature becomes both: hence the identification of Nature with the Great Mother, with all the ambivalence involved, and so too the necessity to (try to) control her fertility. David Abram, in *The Spell of the Sensuous* (Pantheon, 1996), argues that our very language is rooted in the "more-than-human" expressive world: in animal song and speech and even tracks, the murmurings of trees and streams, and the lay of the land itself. All of this was gradually displaced into language, at first oral and imitative, and clearly borrowed from the "more-than-human," but then progressively cut off from its roots, abstracted into a written symbol system whose pronunciation became a phonetic function: so we fell under the spell of spelling, and the land lost its magic. "In order to learn to read we must break the spontaneous participation of our eyes and our ears in the surrounding environment, where they had ceaselessly converged in the synaesthetic encounter with animals, plants, and streams, in order to re-couple those senses on the flat surface of the page. . . ." For yet another account of how we have quite literally "lost our senses," see Morris Berman, *Coming to Our Senses: Body and Spirit in the Hidden History of the West* (Bantam, 1990).

Val Plumwood argues in *Feminism and the Mastery of Nature* (Routledge, 1991) that the environmental crisis has its roots in western culture's *dualizing* of humans and nature, creating a "deep line of fracture between reason and nature," running through many societies and thinkers, and systematically denying and obscuring the human dependence upon and immersion in nature. The results are vast blind spots in our vision of ourselves and the world, through which modern forms of domination have been able to move, and indeed to seem perfectly natural. Other dualisms, such as that between men and women, run parallel and have similar results. See also Stephanie Lahar, "Roots: Rejoining Social and Natural History," in Greta Gaard, ed., *Ecofeminism* (Temple University Press, 1993). For more on this line of thinking, see the discussion of ecofeminism below.

Historian Lynn White, Jr., in a famous essay called "The Historical Roots of Our Ecological Crisis," (*Science* 155 (1967): 1203–7, but very widely reprinted elsewhere) traces the crisis to Christian anthropocentrism. White's piece occasioned an extensive debate: for responses, see the section on ecotheology below. Frederick

Turner, in *Beyond Geography: The Western Spirit Against the Wilderness* (Viking, 1980), works out a somewhat similar theme—that the environmental crisis ultimately has religious roots—in a profound and detailed way. Daniel Quinn's widely read novel *Ishmael* (Bantam, 1992), echoing Shepard, traces the crisis back to the divergence of "Taker" from "Leaver" civilizations.

Political economists trace the crisis to more recent historical factors, especially the rise of capitalism in tandem with modern science. On the transition from medieval to modern political economy and the corresponding "objectification" of nature, see Carolyn Merchant, *The Death of Nature: Women, Ecology, and the Scientific Revolution* (Harper and Row, 1980), and the first part of Morris Berman's *The Reenchantment of Nature* (Cornell University Press, 1981). David Ehrenfeld's *The Arrogance of Humanism* (Oxford, 1978) blames the technocratic mind (this is what Ehrenfeld means by "humanism") and offers religious humility (contra White and Turner) as more like a saving grace. Marxists have their own ecological cross to bear, so to speak—the environmental record of the self-identified Marxist countries has been even more abysmal than capitalism's—but Marxism may still have much to offer as a critical analysis of the nature of capitalism and its characteristic ways of reconstructing and ultimately "commodifying" the world. See the sources cited in the section on environmental politics, below.

Philosophical Reconceptions

So how do we think of ourselves in nature anew? All of the works just cited offer at least some suggestions. Ehrenfeld urges us to recover the suppressed and forgotten Earth-centered strands in the great religious traditions; Berman and others seek in postmodern science the seeds of a "reenchanted" world; Plumwood proposes a many-fronted and practical challenge to dualisms and "centrisms" of all sorts.

The rise of an ecological worldpicture is one impetus toward a systematic reconception of the entire human/nature relationship. An early work pointing the way was Paul Shepard and Daniel

McKinley, eds, *The Subversive Science* (Houghton Mifflin, 1969). Aldo Leopold's *Sand County Almanac* (Oxford University Press, 1949) is now cited as a classic in environmental ethics, indeed as the field's founding text, but it is also a work of larger scope. The argument is that we need to see nature in ways that are not wholly "commercial." Leopold's approach is modest: most of this well-known book is simply a diary of the changes of the year on his slowly regenerating prairie retreat. The larger ethical sense emerges out of that diary, that kind of attentiveness and care. Biographies of Leopold by Susan Flader and Curt Meine give a sense of how he arrived at this point. J. Baird Callicott is one of Leopold's major interpreters in contemporary environmental ethics: see in particular his *In Defense of the Land Ethic* (SUNY Press, 1989).

In Anglo-American thinking, most of the philosophical activity has centered around environmental *ethics*, to which we come presently. But, as in Leopold's and Callicott's work, the turn to ethics has not always been straight to the articulation of the relevant old or new values. Larger frameworks and understandings— of ourselves, of the natural world—have to change also, and often have to change *first*. A number of works often classified as "environmental ethics" actually have this larger concern: for example, Eugene Hargrove, *Foundations of Environmental Ethics* (Prentice-Hall, 1989), more of an aesthetic and historical study than an ethical one; Holmes Rolston III, *Philosophy Gone Wild* (Prometheus Books, 1986); and my own book, *Back to Earth* (Temple University Press, 1994), which tries to open up a sense of the wild possibilities of the more-than-human world.

"Deep ecology" is also in part a philosophical reconception of this sort. A wide variety of different views fit under this label, but the general theme is a strong form of non-anthropocentrism as well as an ecological conception of humans-in-the-environment—a "total field conception," as Norwegian philosopher Arne Naess puts it. For a fairly comprehensive anthology, see George Sessions, *Deep Ecology for the Twenty-First Century* (Boston: Shambala Publishing, 1995). A related philosophical journal is *The Trumpeter* (LightStar Press, Box 5853, Stn B, Victoria, BC V8R 6S8, Canada).

Radical revisions of traditional Western metaphysics have been proposed by twentieth-century Continental (European) philosophers. The German philosopher Martin Heidegger (*Being and Time* (Harper and Row, 1962)) describes human being itself as being a "field of care"—that is, a *field* rather than what others have described as a "skin-encapsulated ego," and a field of *care*, a specific and irreducible kind of engagement. "And being such a field," writes philosopher Neil Evernden, "means more than being a body; it means being-in-the-world, and it implies a different sense of environment." Approach Heidegger through modern interpeters who write with recent environmental concerns in mind, such as Evernden in *The Natural Alien* (University of Toronto Press, 1985 —the above line comes from p. 65), and Bruce Foltz, *Inhabiting the Earth* (Humanities Press, 1995). For an overview of the Heideggerean upshot for environmental ethics, see the discussion in Patti Clayton, *Connection on the Ice* (Temple University Press, 1998).

Other environmental philosophers have been inspired or emboldened by the work of French phenomenologist Maurice Merleau-Ponty, especially *The Phenomenology of Perception* (Humanities Press, 1962), foregrounding the creative role of the body in all perception, which may be linked, in his last and unfinished work (*The Visible and the Invisible* (Northwestern University Press, 1968)), with the larger "flesh" of the world itself. See the discussion in Abram's *The Spell of the Sensuous*, Chapter 2.

Marxists and many others who also consider themselves "radical ecologists" have both challenged and taken up different aspects of deep ecology and Continental environmental philosophy. A kaleidoscopic overview of the whole debate can be found in Michael Zimmerman, *Contesting Earth's Future: Radical Ecology and Postmodernity* (University of California Press, 1994).

Not all philosophical "reconceptions" are by philosophers. Bly's *News of the Universe,* for example, deserves mention again. Edward Abbey's *Desert Solitaire* (Ballantine, 1968) is a fine example of deep ecological narrative. The seemingly inescapable philosophical commitments of the modern age might not be so inescapable beyond the academy and the worlds of government and industry. People may live "deep ecological" lives without the benefit of philosophy, and indeed one focus of the deep ecological movement, as opposed to its academic cousins, is an emphasis on direct action and simplicity of means. Many people are there already. An Eskimo shaman:

> The great sea
> Has sent me adrift,
> It moves me as the weed in a great river,
> Earth and great weather move me,
> Have carried me away,
> And move my inward parts with joy. (Bly, p. 257)

Indigenous Thinking

As Jim Cheney argues in these pages, Native American thinking about environmental issues should not be read as merely another set of answers to the familiar old questions. It is better to walk lightly and with "respect," to try to approach this body of thought and practice without so many preconceptions.

The best introduction to Indigenous worlds is through the words of Indigenous peoples themselves. N. Scott Momaday lyrically explores the interrelationship between the history of the Kiowa, the mythology that grows out of this history, and his own Indian identity and experience—and the relationship of all of these to the land—in *The Way to Rainy Mountain* (University of New Mexico Press, 1988). Most of Momaday's environmental writings have been collected in *Man Made of Words* (St. Martin's, 1996). Simon Ortiz's *Woven Stone* (University of Arizona Press, 1992) exemplifies the rich tapestry of Indigenous life in relationship to a more-than-human world and within the hard reality and struggles of contemporary Indigenous life. Ceremony is explored in a modern context in Leslie Marmon Silko's *Ceremony* (New American Library, 1977) and contemporary Tricksters come alive in Gerald Vizenor's richly comic *The Trickster of Liberty* (University of Minnesota Press, 1988).

An academic journal, *Ayaangwaamizin: The International Journal of Indigenous Philosophy,* under the editorship of Choctaw and Ojibway philosophers, has recently been founded (Bookstore, Lakehead University, 955 Oliver Road, Thunder Bay, ON P7B 5E1 Canada). Other academic works are Sam Gill's *Native American Religions* and *Native American Traditions* (Wadsworth, 1982 and 1983), and John Bierhorst's *The Way of the Earth* (Morrow, 1994), an encyclopedic survey with much cultural detail relevant to environmental philosophy. Three exemplary studies of specific cultures are Richard Nelson's *Make Prayers to the Raven: A Koyukon View of the Northern Forest* (University of Chicago Press, 1983); Keith Basso, *Wisdom Sits in Places: Landscape and Language Among the Western Apache* (University of New Mexico Press, 1996), and Tom Lowenstein, *Ancient Land, Sacred Whale: The Inuit Hunt and Its Rituals* (Farrar, Straus and Giroux, 1993). Calvin Martin's *In the Spirit of the Earth* (Johns Hopkins University Press, 1992) is intriguing in its own right and also offers a useful bibliography. Students and teachers with access to PBS's marvelous series *Millennium: Tribal Wisdom in the Modern Age* may find many of those videos useful as well.

Ecotheology

As noted above, Lynn White and some others place part of the blame for environmental crisis on the Christian tradition. Thinkers within Christianity and other religious traditions, however, are also struggling to find ways to speak to the crisis, and here too there is an outpouring of new writing. Robert Booth Fowler's *The Greening of Protestant Thought* (University of North Carolina Press, 1995) offers an overview. For a traditional Christian response, see Michael Northcott, *The Environment and Christian Ethics* (Cambridge University Press, 1996). A less traditional Christian response is Matthew Fox's *Original Blessing* (Santa Fe, NM: Bear and Company, 1983), which sets out a "creation spirituality" in place of the world-rejecting, sin-oriented tradition—which, Fox argues, has never been the whole story of the Christian tradition anyway.

Two excellent introductions to ecotheology are Roger Gottlieb's collection *This Sacred Earth* (Routledge, 1996) and David Kinsley, *Ecology and Religion* (Prentice-Hall, 1995). Gottlieb pays special attention to Christian, Jewish, and native people's religious discussions; Kinsley concentrates on traditional religions, Asian traditions, and the debate initiated by White. Seyyed Hossein Nasr, *Religion and the Order of Nature* (Oxford University Press, 1996) is a synoptic account of religious response to environmental crisis. Steven Rockefeller and John Elder, editors, *Spirit and Nature* (Beacon Press, 1992) also include a variety of perspectives, including Tibet's Dalai Lama, some Native American religious views, Sallie McFague (a Protestant feminist theologian), and others. PBS offers a videotape with Bill Moyers under the same title.

Most Christian denominations have some environmental organization or agency. Gottlieb cites a number of them at the end of his collection. There is even an Evangelical Environmental Network (10 E. Lancaster Ave, Wynnewood, PA 19096). One umbrella organization is the National Religious Partnership for the Environment (1047 Amsterdam Ave, New York, NY 10025). For Jewish environmental action, contact the Coalition on the Environment and Jewish Life (443 Park Ave South, 11th Flr, New York, NY 10016).

Responding to the environmental crisis has also led to a resurgence of interest in older and Earth-centered religions, such as those labelled "paganism"—though this is not always classified as ecotheology (we don't easily speak of "pagan theology"). See Chapter 16 of the Kinsley text for a brief overview of what is now being called "neopaganism." We also return to some of these themes in the section on ecofeminism, below. For a rather different approach to neo-paganism, see Dolores LaChapelle, *Sacred Land, Sacred Sex* (Durango, CO: Kivaki Press, 1988).

Environmental Ethics

Environmental philosophy in the Anglo-American tradition typically takes the form of environmental *ethics*. A general introduction is Joseph DesJardins, *Environmental Ethics* (Wadsworth Publishing Company, second edition 1997). Widely used anthologies include Susan Armstrong and Richard Botzler, *Environmental Ethics: Divergence and Convergence* (McGraw-Hill, 1993); Christine Pierce and

Donald Vandeveer, *People, Penguins, and Plastic Trees* (second edition, Wadsworth, 1995); and Louis Pojman, *Environmental Ethics* (Jones and Bartlett, 1994).

Common themes of these books are: how far anthropocentrism can take us, and how limiting anthropocentric perspectives (e.g., economics) prove in the end; the sources of indirect and direct obligations to the natural world; and some of the ways that the variety of values at stake might be organized into a single, unified, theoretical framework. Each anthology focuses on different additional issues as well: for example, the cultural, aesthetic, and scientific background in Armstrong and Botzler; environmental policy questions and activism in Vandeveer and Pierce.

A variety of competing theoretical ethical frameworks have been proposed. As Rolston's essay in this collection suggests, ethics has been led to progressively wider and wilder places. We may begin inside anthropocentrism, where, as Rolston argues, there are already powerful incentives to care about nature. See also Bryan Norton, *Toward Unity Among Environmentalists* (Oxford University Press, 1991). A next step is to extend certain kinds of value— rights, or a weighing of welfare—to at least some, perhaps all, nonhuman animals. See Peter Singer, *Animal Liberation* (New York Review of Books Press, second edition, 1990) and Tom Regan, *The Case for Animal Rights* (University of California Press, 1985); there are some other citations in the section on animals below.

For ethical systems that extend value to *all* lifeforms (a view sometimes called "biocentrism"), see Paul Taylor, *Respect for Nature* (Princeton University Press, 1986) and Lawrence Johnson, *A Morally Deep World* (Cambridge University Press, 1991). For ethical systems that extend value to entire ecosystems (a view sometimes called "ecocentrism"), see J. Baird Callicott, *In Defense of the Land Ethic*, cited above. On the possible extension of value to the entire planet, start with my essay "Forms of Gaian Ethics," *Environmental Ethics* 9:3 (1987). "Deep ecology," as in the volume edited by George Sessions cited above, also offers a (less professionalized) form of (something like) ecocentrism.

For an extensive overview of the entire process, see the many works of Holmes Rolston, especially his *Environmental Ethics: Du-*

ties to and Values in the Natural World (Temple University Press, 1988), and *Philosophy Gone Wild* (Prometheus, 1986). See also Peter Singer, *The Expanding Circle* (NY: Farrar, Straus, and Giroux, 1981). Tom Birch's "Moral Considerability and Universal Consideration," *Environmental Ethics* 15:4 (1993) proposes a more direct and mindful, less hierarchical form of "universal consideration."

Journals in the field include *Environmental Ethics* (Department of Philosophy, University of North Texas, Denton, TX 76203-0980), *Environmental Values* (Whitehorse Press, 10 High Street, Knapwell, Cambridge, CB3 8NR, UK), and *Ethics and the Environment* (JAI Press, Inc., 55 Old Post Road, #2, Greenwich, CT 06836). The International Society for Environmental Ethics (c/o Laura Westra, Department of Philosophy, University of Windsor, Windsor, ON N9B 3P4, Canada) publishes a newsletter and maintains a website through the Center for Environmental Philosophy at the University of North Texas: the address is <http://www.cep.unt.edu/ISEE.html>. Also accessible through the UNT website is a Master Bibliography in Environmental Ethics.

Ecofeminism

Philosophers such as Val Plumwood (in *Feminism and the Mastery of Nature*, cited above, and in her essay for this collection) and Karen Warren ("The Power and Promise of Ecological Feminism," *Environmental Ethics* 12:2 (1990) and widely anthologized elsewhere) explore the connections between the oppression of women and our treatment of other animals and nature as a whole. That many kinds of oppression are linked in this way, and that the oppression of women is the most fundamental or one of the most fundamental of these linked oppressions, is a kind of view now known as *ecofeminism*.

One strand of ecofeminism aims to recover what are said to be deeper female connections to nature: a sense for nature's cycles, life and death, love and care as primary world-orientations rather than what are said to be characteristically "masculine" abstraction and distance. Some "cultural ecofeminists" argue that these connections need to be articulated and celebrated in new ways. Others want to recover older forms of life in which, they suggest, these connections already were richly articulated. As one "Wiccan" ecofeminist presents it, for example, even witchcraft, understood for once without all the derogation and uneasiness that we have learned to associate with it, is really no more or less than the celebration and practice of intimacy with the Earth: with her cycles and powers, mysterious and deep as we are only now beginning to (re?)discover that they are. See Starhawk, *Dreaming the Dark: Magic, Sex, and Power* (Beacon Press, 1982) and *Truth or Dare* (Harper and Row, 1987). This kind of ecofeminism lends itself directly to practice: for example, to the kinds of seasonal celebrations and remembrances noted above (under "Coming to our senses"). All of Starhawk's books, for example, offer seasonal and other rituals in a Wiccan key.

Other feminists criticize cultural ecofeminism as what Plumwood calls "the feminism of uncritical reversal," which by embracing stereotypical female traits perpetuates women's oppression—not to mention gender dichotomizing—in a new and more subtle form. Plumwood argues that the task instead is to identify and challenge the dualisms and hierarchies that underlie this and other forms of oppression. This more political and critical kind of ecofeminism also lends itself directly to practice, though a different kind of practice: like all centrisms, as Plumwood points out, anthropo-centrism can be resisted and reversed by deepening understanding of nature's own dynamics, an appreciation for ecological uncertainty and fragility, and "the cultivation or recovery of ways of seeing beings in nature in mind-inhabited ways, as other centers of needs and striving. . . ."

An additional contribution of ecofeminism is to awaken an awareness of how much on-the-ground environmental protection work around the world is done by women. See Vandana Shiva, *Staying Alive: Women, Ecology, and Development* (London: Zed Books, 1988); and Sally Sontheimer, ed., *Women and the Environment* (NY: Monthly Review Press, 1991).

Classic ecofeminist works include Susan Griffin's *Women and Nature* (Harper and Row, 1978) and Carolyn Merchant, *The Death of Nature: Women, Ecology, and the Scientific Revolution,* cited above. A helpful recent philosophical anthology is Greta Gaard, *Ecofeminism: Women, Animals, Nature,* cited above.

Animals

Other animals surround us almost all the time: birds in the skies; insects underfoot and buzzing in the air; dogs and cats. Historically, humans have almost always lived in community with other creatures—from cormorants to weasels to all manner of farm animals, camels, horses, even ants and snakes—as Mary Midgley has pointed out (see her *Animals and Why They Matter* (University of Georgia Press, 1984)). This relationship already has an implicit ethics. Writer and animal trainer Vicki Hearne articulates moral worlds that humans and certain domestic animals co-create in her book *Adam's Task* (Knopf, 1986).

On the extension of traditional individualistic ethical frameworks to other animals, see the works of Singer and Regan cited above (in "Environmental ethics"). For an ecofeminist view, see Carol Adams, *The Sexual Politics of Meat* (Continuum, 1990). On the sometimes troubled and unclear relation between a Singer/Regan type of ethics and environmental ethics, see Eugene Hargrove, ed., *The Animal Rights/Environmental Ethics Debate* (SUNY Press, 1992). For philosophically and ethically minded exploration of human-ape relations, see Paola Cavalieri and Peter Singer, eds., *The Great Ape Project: Equality Beyond Humanity* (St Martin's, 1993); Eugene Linden, *Silent Partners: The Legacy of the Ape Language Experiments* (NY: Times Books, 1986); and Sy Montgomery, *Walking with the Great Apes* (Houghton Mifflin, 1991). A striking philosophically minded meditation on human-animal relations generally is Barbara Noske, *Humans and Other Animals* (London: Pluto Press, 1989). Recall also Nollman's *Animal Dreaming,* cited above.

Singer's book ends with meat-free recipes (a philosophical argument ending with recipes—surely a first in the field!). John Robbins' *Diet for a New America* (Walpole, NH: Stillpoint Publishing, 1987) lays out ethical, dietary, and ecological arguments for veganism (i.e., eating no animal products at all). Vegetarian and vegan cookbooks can be found in any alternative foodstore and almost any bookstore. On vegetarian diets for children, see Victoria Moran, "Getting Started as a Vegetarian Family," *Mothering*, Fall 1996. The traditional "Four Food Groups" (actually promoted in 1956 by the FDA at the instigation of the meat and dairy industries) have been revised into a "food pyramid" de-emphasizing meat and dairy products. Contacts include the Vegetarian Resource Group (PO Box 1463, Baltimore, MD 21203); the North American Vegetarian Society (PO Box 52, Dolgeville, NY 13329); the American Vegan Society (501 Old Harding Highway, Malaga, NJ 08328). Many cities have flourishing vegetarian groups: watch the bulletin boards at alternative food stores and co-ops.

There are many pro-animal organizations working on other animal issues, from drug testing to to zoos: from fairly radical organizations like PETA (People for the Ethical Treatement of Animals, PO Box 42516, Washington, DC 20015) to Humane Societies (Society for the Prevention of Cruelty to Animals) everywhere. There is a good bit of debate and even animosity between such groups (as between a number of the other groups mentioned in this essay), springing partly from fairly deep philosophical differences about the sources of the current crisis and the appropriate responses to it: so look into any organization before you join it— and take a look at its critics, too! Most of these organizations publish helpful newsletters and action alerts.

As far as I know, there are no worked-out ethical defenses of factory farming; it is hard to escape the conclusion that it is a practice sustained by silent collusion, by the wish "not to know." There are, however, defenses of the use of animals in research (mostly on utilitarian grounds), and a variety of defenses of hunting. Some radical ecologists defend hunting as a form of ecological participation: see Paul Shepard, "A Post-Historic Primitivism," in Max Oelschlager, ed., *The Wilderness Condition* (Sierra Club, 1992). Hunters are sometimes the best informed citizens on ecological issues, and many wildlife organizations have been strong advocates for environmental protection. Hunting itself has an ethic: see Jim Posewitz's *Beyond Fair Play: The Ethic and Tradition of Hunting*

(Falcon Press, 1994), a text widely used in hunter education. Those committed to animal rights, on the other hand, argue that hunting inevitably is an unnecessary and unethical form of cruelty. A striking meditation on the whole matter is Ted Kerasote's *Bloodties: Nature, Culture, and the Hunt* (Random House, 1993). Contra Posewitz, see Brian Luke's article "A Critical Analysis of Hunters' Ethics," *Environmental Ethics* 19:1 (1997)) which argues that the "Sportsman's Code" promoted by Posewitz and others, if taken to its logical conclusion, implies that hunting itself is ethically wrong.

Other organizations concerned for animals range from Greenpeace (1436 U. Street NW, Washington, DC 20009) which is inspired by the Quaker practice of "witnessing" and dedicated to non-violent direct action to save sea animals, wolves and other endangered creatures (and also a larger environmental agenda, like stopping nuclear tests and toxic pollution), to the National Audubon Society, cited above, and Biodiversity Legal Foundation (files lawsuits on behalf of endangered species: PO Box 18327, Boulder, CO 80308); Earth Island Institute (300 Broadway, Suite 28, San Francisco, CA 94133); Defenders of Wildlife (1244 19th Street NW, Washington, DC 20036); and a number of local organizations focusing on specific species.

Environmental Politics

As noted above, some thinkers lay the blame for our ecological destruction on capitalism and the political and scientific patterns of thinking (distanced, abstract, mechanistic, and hence exploitative) that come with it. Economic themes, and economic change, therefore seem more essential than strictly philosophical or ethical change (Marxists tend to see intellectual conceptions in general as derivative from economic forms of social organziation). For a Marxist approach to environmentalism, see Howard Parsons, *Marx and Engels on Ecology* (Westport, CT: Greenwood Press, 1977) and Ted Benton, ed., *The Greening of Marxism* (NY: Guilford Press, 1996). On the other hand, Marxist discussions of environmental crisis are often (and justly) fairly defensive, since Marxism itself has many elements of a dominating attitude toward nature, and the environmental record of Marxist countries borders on the catastrophic. See for example Andrew McLaughlin, "Marxism and the Mastery of Nature," in Roger Gottlieb, ed., *Radical Philosophy: Tradition, Counter-Tradition, and Politics* (Temple University Press, 1993). Part III of Benton's collection ("The Second Contradiction of Capitalism") is a notable exception.

For an attempt at a systematic "green" political theory, see Robert Goodin, *Green Political Theory* (Cambridge: Polity Press, 1992). Robyn Eckersley, *Environmentalism and Political Theory* (SUNY Press, 1992) is a useful survey. Philosopher Mark Sagoff has made a systematic project of resisting the reduction of political choices to economic terms, especially as relating to environmental issues: see his *The Economy of the Earth* (Cambridge University Press, 1988). Further debate on environmental economics can be found in Pierce and Vandeveer, *The Environmental Ethics and Policy Book* (Wadsworth, 1994). In *For the Common Good* (Beacon Press, 1989), Herman Daly and John Cobb outline a new vision of environmental economics. Rigorously and srikingly argued is Richard and Val Routley's (later Richard Sylvan and Val Plumwood) essay "Nuclear Power: Some Ethical and Social Dimensions," in Tom Regan and Donald Vandeveer, *And Justice for All: New Introductory Essays in Philosophy and Public Policy* (Totowa, NJ: Rowman and Littlefield, 1982). An environmental ethics and public policy website is <http://divweb.harvard.edu/csvpl/ee/>.

For a survey of environmental ethics from the point of view of economic and social justice concerns, see Peter Wenz, *Environmental Justice* (SUNY, 1988). Carolyn Merchant reviews some of the standard ethical and political views from a more radical political and economic point of view in her *Radical Ecology* (Routledge, 1992).

For a striking work by a contemporary politician, see Albert Gore, *Earth in the Balance*, cited above. Gore surveys the current environmental crisis, offers a deep diagnosis of it, and proposes a range of political initiatives in response. There are not too many books by vice presidents of the United States that speak of themes like "dysfunctional civilization," "we are what we use," and "environmentalism of the spirit."

Gore disappointed many of his environmentalist supporters once taking office. Perhaps he feels constrained by the politics of the possible, especially in an age that is perceived by politicians, anyway, as an age of environmental backlash. If so, maybe he needs some help. All of us can write letters and make phone calls. Some of the major lobbying organizations are already mentioned in this essay: there are also Global Response (PO Box 7490, Boulder, CO 80306), the League of Conservation Voters (1707 L Street NW, Suite 550, Washington, DC 20036); Public Citizen (2000 P Street NW, Suite 610, Washington, DC 20036; this is Ralph Nader's umbrella organization. Anti-toxics, anti-nuclear, pro-product safety and campaign finance reform: learn how all of these issues are connected); and Rainforest Action Network (450 Sansome, Suite 700, San Francisco, CA 94111). The National Wildlife Federation maintains a World Wide Web site at <http:www.nwf.org/nwf> with weekly updates on pending environmental legislation.

It is also worth noting that environmental commitments are much broader and stronger in the general population than it might seem from those who get the most attention. Sagoff argues that it is not environmentalists who are "elitists" and unrepresentative, as is often claimed, but rather their corporate and ideological opponents. See his "'I am no Greenpeacer, But . . .': Environmentalism, Risk Communication, and the Lower Middle Class," in W. Michael Hoffman, et. al., *Business, Ethics, and the Environment* (NY: Quorum Books, 1990). Another revealing source is Willet Kempton, James Boster, and Jennifer Hartley, *Environmental Values in American Culture* (MIT Press, 1995).

Remember too that there are organized routes to social change that do not run through Washington. Earth First! (PO Box 1415, Eugene, OR 97440) advocates non-violent direct action to stop logging, dams, and other destruction: sit-ins, road blockades, demonstrations on-site. EF! in fact was started by former lobbyists in search of another and more direct route. It has had a stormy history, vilified in mainstream media and pursued by FBI for tree-spiking, for example, and other "eco-sabotage." See John Davis, ed., *The Earth First! Reader* (Salt Lake City, Utah: Peregrine Smith Books, 1991). In a very different corner there is the Nature Conservancy (1815 N. Lynn St, Arlington, VA 22209), whose strategy is to quietly buy up land to protect and preserve it, sometimes transferring it later to state or federal authority and sometimes managing the land itself.

Classic philosophical works in the area of environmental policitcs and the law are Christopher Stone, *Should Trees Have Standing? Toward Legal Rights for Natural Objects* (Los Altos, CA: Wm Kaufmann, 1974; reissued by Oceana Publications (Dobbs Ferry, NY, 1996) with a number of Stone's other essays included and a review of progress since the original publication); and Stone, *The Gnat is Older Than Man: Global Environment and Human Agenda* , cited above. See also, again, Bryan Norton, *Toward Unity Among Environmentalists.*

Wilderness and the City

Environmentalism is sometimes reduced, even by environmentalists, to wilderness protection. In fact, though, although we certainly care for and love wilderness in a special way, it is by no means our only concern. All of the contributors to this volume, and most environmental thinkers generally, are concerned for the *whole* Earth and the quality of our *whole* lives, which, as Rolston notes, predominantly include urban and rural as well as wild environments.

There is even some skepticism concerning the place of wilderness in the whole picture, and the sometimes anti-environmental uses, both philosophical and practical, to which the notion and practice of wilderness have been put. See Tom Birch, "The Incarceration of Wilderness," *Environmental Ethics* 12:1 (1990). Rolston and J. Baird Callicott had an extensive exchange in *The Environmental Professional* on these themes, reprinted in Lori Gruen and Dale Jamieson, *Reflecting on Nature* (Oxford University Press, 1994). For classic histories of the concept of wilderness, see Roderick Nash, *Wilderness and the American Mind* (Yale University Press, third edition, 1982) and Max Oelschlager, *The Idea of Wilderness* (Yale University Press, 1990).

Back in the places we live, meanwhile, the philosopher-farmer-poet Wendell Berry writes that "The question that must be addressed . . . is not how to care for the planet, but how to care for each of the planet's millions of human and natural neighborhoods, each of its millions of small pieces and parcels of land, each one of which is in some precious way different from all the others." (Wendell Berry, *What are People For?* (San Francisco: North Point Press, 1990), p. 200.) See also Sara Stein's appropriately titled *Noah's Garden: Restoring the Ecology of Our Own Backyards* (Houghton Mifflin, 1993). Many thinkers are now looking at "mixed landscapes" and urban ecology with environmental concerns in mind: for an introduction, see Chapter 6 of my *Back to Earth*, and Tony Hiss, *The Experience of Place* (Vintage, 1990). The work of Christopher Alexander and his colleagues, most notably *A Pattern Language* (Oxford University Press, 1977) has been widely influential. *A Pattern Language* is a guide to everything from city design to how to frame a window: in each case the concern is for openness to both the human and other-than-human depth of things. A wonderful little newsletter linking work like Alexander's, Murray Schafer's study of "soundscapes" around the world, and others, including some philosophers and phenomenologists, is the *Environmental and Architectural Phenomenology Newsletter* (c/o David Seamon, Architecture Department, Kansas State University, Manhattan, KS 66506).

Bioregionalism and Permaculture

Most bookstores carry little books with titles like *50 Simple Things You Can do to Save the Planet* (Earthworks Press, 1989), such as carpooling, turning down the thermostat, taking care with housepaints, and so on. There is good, detailed data here, especially on the cumulative environmental effects of taken-for-granted daily habits and household items. See also Earthworks' *Student Environmental Action Guide* (1991). Books such as *50 Simple Things* in turn have spawned more challenging lists of environmentally responsible acts, like not driving or flying at all, not having any children, not eating anything that comes in a package, and so on—this from the *Earth Island Journal*, published by the Earth Island Institute (address above), the point being that improving the environment is not as simple or as effortless as we are sometimes led to believe.

It is better to remember that not only can the rest of the world's peoples not live like the hyperindustrial nations in any kind of sustainable way, but even *we* cannot live as we do in any kind of sustainable way. So why are we still trying to get away with it? How long do we expect to do so? A serious question.

The problem with these approaches is that they tend to be negative: don't do this, stop doing that. What we really need is a vision of what we *should* do, how we *should* live, so that our consumptive and destructive habits might even wither away without being noticed. This is the aim of new movements like bioregionalism and permaculture. Bioregionalists are people who commit themselves to paying careful attention to local places, local ecosystems, looked at ecologically rather than, as usual, socially and politically. Bioregionalists locate themselves by watershed, plant communities, and so on, rather than by arbitrary human-imposed lines in the landscape. Some commit themselves to stay in a place for life: this is called "reinhabitation." See Kirkpatrick Sale, *Dwellers in the Land: The Bioregional Vision* (Sierra Club, 1985), and *Home! A Bioregional Reader*, ed. Van Andruss et al. (New Society Publishers, 1990). On the link back to ethics, see Jim Cheney, "Postmodern Environmental Ethics: Ethics as Bioregional Narrative," in *Environmental Ethics* 11:2 (1989).

Sometimes it seems that we know more about the *global* environmental crisis than we do about what is going on down the street. As I note in my essay "Is it too late?," one effect is that we feel disempowered. No one feels that they can do "enough" on the global scale. In fact, *especially* next door, there's much to do. One example: one of the private water-monitoring groups in North Carolina, concerned for the health of the estuaries and bays of the state, is now looking upstream and forming a statewide corps of "Streamkeepers" who will regularly walk, watch, and test the waters of all the little streams that flow into the rivers that flow into the bays. My family and I have our own two miles or so of local stream to get to know and to watch. There may be similar organizations in your area. For a national directory of water-quality monitoring opportunities, contact The Izaak Walton League of America, Save Our Streams Program, 707 Conservation Lane, Gaithersburg, MD 20878. For other similar local opportunities, contact your local Sierra Club or Audubon Society affiliate or other local environmental coalitions.

"Permaculture" names a broad movement looking to devise and recover cooperative and sustainable ways of living with the land: forms of agriculture that restore and enrich the soil, rather than impoverish and destroy it; technologies that work with nature rather than against it; and the forms of social organization that (we are increasingly recognizing) must accompany and support these agricultures and technologies. On agriculture, see for example the works of Wes Jackson (his classic collection, *Meeting the Expectations of the Land* (North Point Press, 1984), and a new collection, just as good, *Rooted in the Land: Essays on Community and Place* (Yale University Press, 1996)) and Wendell Berry (*The Unsettling of America* (Sierra Club, 1986) and *Home Economics* (San Francisco: North Point Press, 1987)). Jim Mollison's *Permaculture: A Practical Guide for a Sustainable Future* (Island Press, 1990) is an all-purpose practical compendium, and no resource library would be complete without the *Whole Earth Catalogs*, especially the *Whole Earth Millennium Catalog* (Harper SanFrancisco, 1994). There are local permaculture associations in many areas.

That all of this also has a poetry—that it is not just another after-the-fact, grudging acceptance of a new set of necessities, but a deeply fulfilling and *right* way of life—is abundantly clear in the work of poet-philosopher Gary Snyder. See his *The Practice of the Wild* (San Francisco: North Point Press, 1990) and many volumes of poetry: a good place to start is *No Nature: New and Selected Poems* (Pantheon, 1992). And with the politics of this poet-laureate of bioregionalism, we find a good place to end:

> I pledge allegiance to the soil
>   of Turtle Island
> and to the beings who thereon dwell
>   one ecosystem
>   in diversity
>   under the sun
> with joyful interpenetration for all.

A. W.